COLIN CHAPMAN
LOTUS ENGINEERING

NVW 239C

OSPREY AUTOMOTIVE

COLIN CHAPMAN
LOTUS ENGINEERING

HUGH HASKELL

FOREWORD BY STIRLING MOSS OBE

OSPREY
AUTOMOTIVE

For Carole

HALF TITLE **Colin Chapman driving the author's special Lotus 7 (known as the 7/20) at Silverstone on 11th August 1962 in the Six Hour Relay Race. This car was fitted with a full Formula Junior style suspension system grafted from the Lotus Mark 20. While, at the time it was built, the car was entirely a private venture, it no doubt contributed something to the Type 37 which was built after the author had left the company. Four cars similar to this one were built in the two years that followed; the original, having been beautifully restored in California, is at the time of writing in the hands of an enthusiast in Tokyo.**

TITLE PAGE **Bengt Soderstrom flying low in his Lotus Cortina, 1966.**

FRONT OF JACKET **1957 Lotus Seven Series I. Do not be misled: this is not the author's personalised registration plate, the Seven belongs to Caterham Cars. (Photo: Andrew Morland)**

BACK OF JACKET **A day at the races. (Photo: Behram Kapadia)**

First published in 1993 by Osprey,
an imprint of Reed Consumer Books Limited,
Michelin House, 81 Fulham Road,
London SW3 6RB
and Auckland, Melbourne, Singapore and Toronto

Cataloguing in Publication Data for this title is available from the British Library

ISBN 1 85532 376 1

Project Editor Shaun Barrington
Page design Gwyn Lewis
Filmset by Tradespools Limited
Printed in Great Britain by BAS Printers Limited, Over Wallop, Hampshire

Contents

Acknowledgements

MY GRATEFUL THANKS are due to all those who have helped me in the preparation of this book: Albert Adams, with whom I worked many years ago, and who helped with reminiscences, contacts and illustrations. Shaun Campbell, Editor of **Autocar and Motor**, who helped with contacts. Clive Chapman, who helped with his memories of his father and facilitated my access to others in Team Lotus. Charlie Chester, through whose help I found Doug Springham. Nick Collins, without whom this book would not have been written, or would have been less readable. David Cook, for help on his creation, the Shadow aircraft. John Cooper, who gave me a warm welcome and provided much interesting information on the early days. Mike Costin, by whom I was recruited into Lotus in 1960 and whose recollections helped to complete much background. Ian Doble, whom I met only during the writing of the book, when he enthusiastically shared his memories. Jeremy Edwards, a college friend who gave great help in collating information and was supportive in many ways. Andrew Ferguson, Manager of Team Lotus in the 1960s. Andrew was a great help in the supply of information and in making contacts. Mauro Forghieri, for his kind responses to my written enquiries. Colin Gething, who gave his time to reminisce and lent his personal albums of photographs. Ken Gregory, for many years Stirling Moss' manager, who helped to clarify details from the past. Gordon Jarvis, from whom I bought a car and came by some useful background. Denis Jenkinson: I was most flattered to receive help from someone who has been an idol for many years. Gwen Kell, for her kind help in providing access to Lotus' press files. Hugh Kemp and Simon Wood, who helped on engine history. Pam Leggate, who managed somehow to pull together the engagements of several busy people and smoothed things along most effectively. Keith & Wendy Hamblin, two lifetime friends who provided copious support and reference information. Dr Lucy Henry, for psychological insight. Stirling Moss, another idol from my youth who gave time and help well beyond expectations, including the most kind Foreword. Gordon Murray, McLaren's Chief Designer and fan of Colin Chapman, who found time during a World Championship season (one of four now) to help most generously. Patrick Peal, for his help on the Microlight Aircraft background. Tony Rudd, who talked freely not only about his time at Lotus, but allowed me also a view of Chapman's involvement with BRM. Steve Sanville, who worked on transmissions and was later in charge of engine development. His contribution was most valuable. Doug Springham, who responded to an appeal put out on Charlie Chester's radio show and produced some marvellous insight into the young Colin Chapman. Peter Thomas, once a neighbour, who gave valuable help with research, as well as advice. Tony Thacker, who first suggested I might write something (although he would not perhaps have guessed what would eventually come of it). Peter Wright: his analytical mind and thorough knowledge of ground effect were a great support in setting the scene for this subject. Occupiers of the Ex-Lotus factory at Cheshunt: Paul Burdett and Derek Henderson of Biccotest Limited, makers of hi-tech test equipment. Ron and Ian Harding, Electrocircuits (P.C.) Ltd, makers of printed circuits. John Offer, Power Development Ltd, makers of varistors. Nick Toper and Allan Roberts of Omega, makers of furniture. Finally, those who helped this, my first venture into writing, by showing an encouraging interest throughout and making me feel at times like a real author. These are too numerous to mention and some of them are even unknown to me. I thank them all.

Foreword

GENIUS IS A very special word, and in an engineering context, Colin Chapman was a very special man. He has probably given our sport more innovative ideas than any other engineer. He was a very competent driver, a good team manager and a difficult man. His cars were probably the best in their era, but they did require an understanding driver with considerable experience. When he was joined by Jimmy Clark, one of the world's greatest drivers, their partnership was as formidable as any, ever.

This book has told me an awful lot about this unique man. It is written in an easy to read manner, where technical terms and methods are explained in a way that one can understand and yet without ever feeling foolish for not knowing.

I wish I had known the man better and had taken the opportunity to discuss his technical ideas, hopes and ideals, but, as I didn't, this book has, happily, given me the chance to enjoy finding out about genius.

Stirling Moss

Introduction

AT SCHOOL AGE I was interested only in aeromodelling and otherwise showed my 'mechanical interest' by stealing watches my father had for repair to take to pieces under my bed.

By university age I had caught the car bug and while studying for my B.Sc. in engineering, would sit at the back of class reading the weekly magazine *Autosport* and planning rallies for the university motor club. After some years of such full time education (to which I was clearly not committed), plus a further year of part time study while working in a garage, I decided that I wanted to do something more interesting with my life.

So, early in 1960 I hired a typewriter and wrote to every sports/racing car company I could think of. By the time of the next major race at Brands Hatch I'd still not had any reply from Lotus, so I went to the practice day and accosted Colin Chapman. Perhaps his first words to me tell us something about him. When I said I was looking for a job he said 'Well we don't really want enthusiasts you know – they're always dreaming about the next race rather than getting the job right'. Fortunately for me I had just overheard him berating a mechanic for getting a wrong selection of gears in the transaxle of John Surtees' Lotus 18 F1, which allowed me to reply that at least enthusiasts wouldn't make that sort of mistake (not true, perhaps, but it got me a formal interview at Cheshunt). My letters had meanwhile led to other interviews so that I had in total four job offers including one from BRM and one from Lotus. Not knowing how to choose, I mentally tossed a coin and went to BRM. As a result of this mistake I found myself writing to Lotus to ask if I could go there after all. This was the start of four years of fun, hard work, some frustration, but above all, learning.

Working with Chapman and Mike Costin (who was my immediate boss to start with) was an initiation without equal. During the time I spent at Lotus I learned to see the fundamentals of a problem – something which I believe is essential for an engineer and something at which Chapman really excelled. In fact, I found that after a short period there I progressed from being a beginner, whose work had to be guided, to feeling that I was trusted simply to get on with it. I quickly felt that Chapman recognised that, within the limits of my smaller domain, I had the same approach to the work. I have always remembered with gratitude this progressive increase of confidence.

In my career as an engineer ever since I have felt the benefit of the guiding hand I had from Chapman and I now feel ready and able to write this book about one of the very few internationally known figures to have featured in my life. To be scrupulously fair I must say that during that period I had a similar level of guidance from Mike Costin, with whom I worked more closely, if over a shorter period.

The idea for this book arose almost by accident and developed over more than a year before any serious work started. The book is a review by a professional engineer of the life and work of another engineer – one who achieved prominence, fame and some notoriety. It sets out to look primarily at the engineering innovations made by Colin Chapman and to show why they were good ideas and why, or in what way, they were original. It covers the early days when Chapman's potential first began to show, his first clashes with the racing establishment, and attempts to lead the reader through the evolution of Chapman's creativity up to his untimely death. In order to make the book readable to those both with and without an engineer's training, I have tried to blend the language used without, I hope, losing clarity. I have therefore given non-technical, or semi-technical explanations of some of the basic principles of mechanical engineering which can be skipped by engineer readers (they will no doubt however, prefer to read them so as to criticise them).

Mentioning the fact that I was writing about Colin Chapman brought a very wide range of responses. An accountant friend said: 'I should like to read about his financial dealings', and from a computer professional: 'I'm not sure I remember who he was', and from a work colleague: 'Won't you need his permission?'. But from motoring enthusiasts everywhere the response was of great positive interest, which I found most encouraging.

One of the things that increases the difficulty of writing a book such as this is that Chapman naturally surrounded himself with other creative people. This means that quite probably some of the ideas I shall talk about as Chapman's in fact originated with any one of several brilliant people who worked with him. In addition, Chapman's competitors were also very creative, which further compounds the difficulty of correctly attributing origin. If I tread on any toes I'm sorry.

Hugh Haskell

Author's Note
Engineering, like any science, must be learnt; but equally clearly, those with the right flair for it in their mental make-up can apply what they have learnt much better than others and can leap forward from it. Add the magic of creativity and you have the potential for greatness.

Some engineering principles will be explained to lay a little groundwork and to make clearer why Chapman was so good at applying the principles when examples are given.

This section is therefore intended mainly for those without engineering training.

Stress is not only what happens under pressure from work or between people, but is perhaps one of the most important elements to be grasped by the would-be engineer. It is the amount of force or load applied to a part in proportion to the amount of material in the part (and also the way the material is arranged).

Engineers divide stress into a number of different types according to the way it tends to distort the material.

Tensile stress is a stretching of the material, and is what happens to a string when we pull on it. *Compressive stress* is negative tensile stress and clearly the string cannot withstand it. It is what the soles of your shoes experience when you stand, or the legs of your chair (or your behind) when you sit.

Bending stress is that which you apply to a lever when you pull it.

Torsional stress is the result of a twisting action and is what the shank of a corkscrew feels when it is screwed into a cork.

Shear stress is what the cork feels as it is pulled out, as its outside surface is dragged downwards by friction with the bottle neck and its centre is pulled upwards by the corkscrew.

From these definitions the pioneers in engineering as a science built the total structure within which most mechanical engineers work.

One of the first principles learnt in connection with the design of engineering structures is that of triangulation. This is, if you like, a 'scientific' recognition of the fact that if you join three straight links or rods together end to end to form a triangle they can only form one triangle.

Contrast this with the fact that if you similarly join four links together they can form any number of quadrilaterals as they pivot about their connection points. To make a quadrilateral rigid you must fix the corners rigidly, whereas a triangle is rigid even if the corners are not.

In engineering terms we can say that a quadrilateral (or polygon) will depend upon the bending strength of the corners – and the sides – for its rigid-

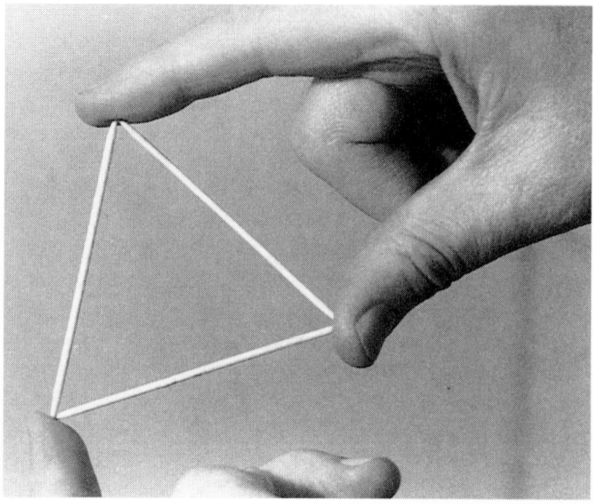

The three stick test: Arranging three sticks as shown demonstrates the effect of triangulation. Such a simple structure gives extraordinary benefits. The reader is encouraged to try this test personally and not just rely on a description. The rigidity resulting has to be experienced to be appreciated.

The six stick test: This is an extension of the three stick test into three dimensions. It is much more difficult to set up, being best done with a helper. Again the stiffness of the pyramid so formed is remarkable.

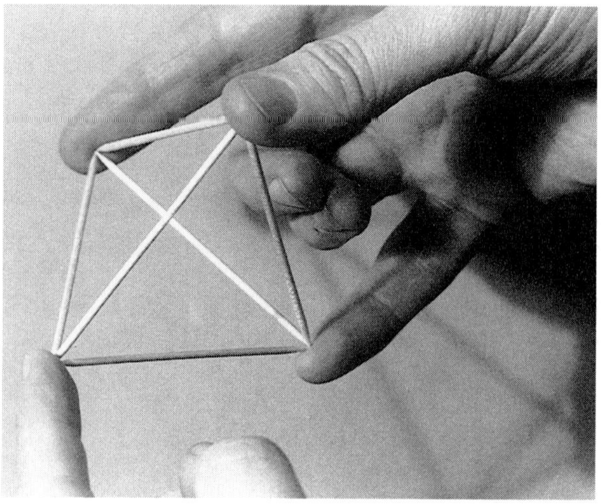

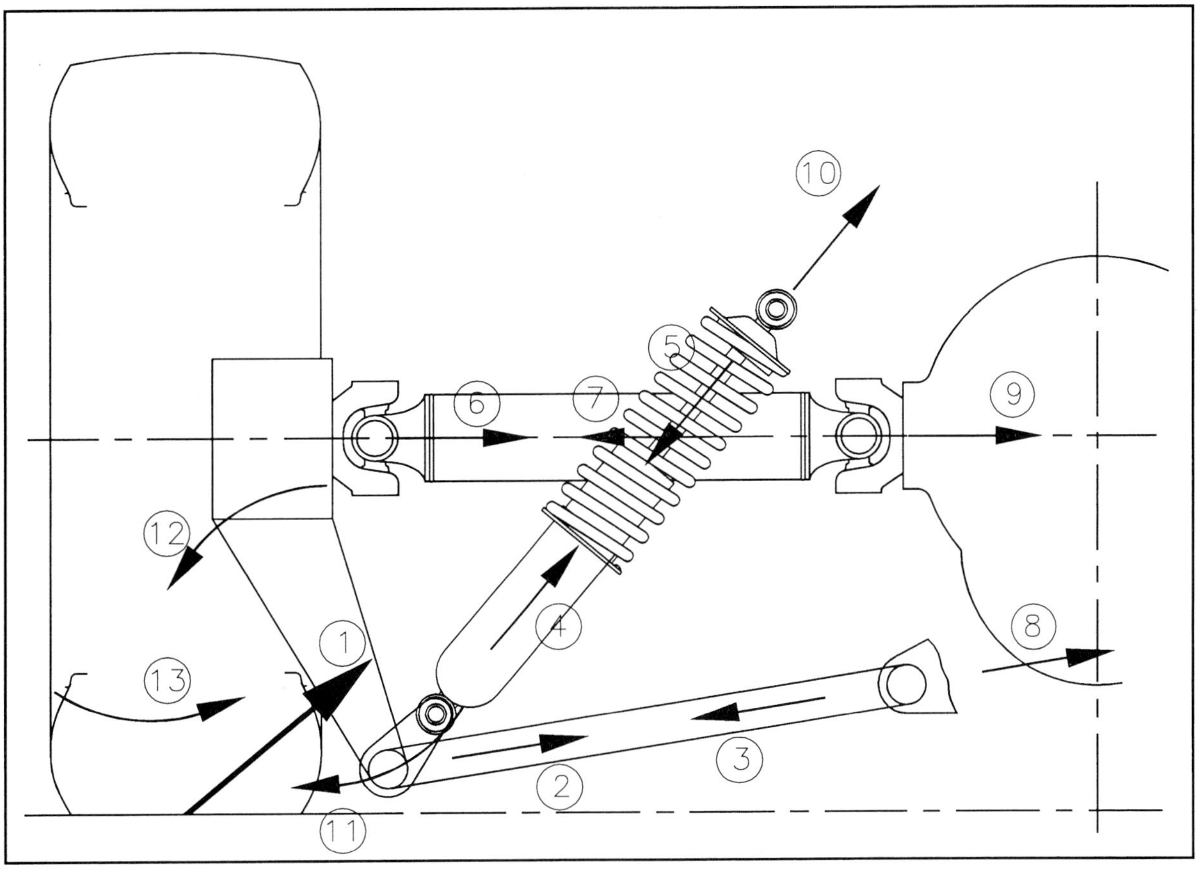

Some of the forces on a suspension system: The large arrow (1) represents the force acting on the tyre where it contacts the road. It is the combination of the weight of the car (vertical) and the horizontal force generated by cornering – turning towards the right in the picture. Arrows (2),(4) and (6) show the forces in the lower wishbone, spring/damper unit and drive shaft respectively, which result from force (1). The opposing arrows (3), (5) and (7) remind us that these forces are compressive. They result in turn in thrust loads being applied to the chassis (8) and (10) and to the gearbox (9). Forces (11) and (12) are those felt by the suspension upright and are curved to illustrate that they result in bending loads on the upright. Similarly the initial load (1) causes bending (13) in the wheel.

ity but a triangle depends only on the tensile or compressive strength of the sides.

Here it is perhaps important to link the ideas of stress outlined above to the equally significant related idea of stiffness. Since any load on a component will cause a stress, the stress will cause a deformation.

It is necessary to separate the notion of *stiffness* from that of *strength.* A wooden ruler held so that half its length protrudes over the edge of a desk will support two pounds or more but will bend noticeably under a much lesser load. A piece of blackboard chalk on the other hand, could be loaded to breaking point without any bending becoming noticeable. The ruler is *stronger* but the chalk is *stiffer.*

Of course, if the ruler is placed edge-on it then becomes stiffer, which is an introduction to the way the stiffness and strength of a part depend upon the way the material is arranged, as well as how much there is. We are all used to the idea of stiff materials which break without deforming and we call them brittle. It is the aim of the metallurgists who make steels and other structural materials to achieve high stiffness without associated brittleness.

Designers of racing cars are always striving for the necessary stiffness to allow the structure to hold its shape while having the minimum of weight. This means that efficient use of the material is at a premium. Going back to our quadrilateral of links for a moment, and bearing in mind the principle of efficient use of material, the best way of fixing the quadrilateral is to put in a diagonal which triangulates it. This method is not always practical as there may be some essential part of the structure we are making which occupies the space (the driver for example).

Many applications in car design demand the highest possible torsional stiffness – the way a part or a structure resists twisting. If you hold in your hands the cardboard tube from the middle of a toilet roll and try to twist it (no, not too hard!) you can sense that it has quite surprising stiffness. In fact it will hardly show any torsional movement, one end relative to the other, up to the point where it collapses altogether. This is because, as engineers can prove mathematically, a round tube represents the best arrangement of material to resist torsion.

With any other form you will find considerable yielding before actual failure – try with a long narrow square section cardboard box if you can find one (perhaps the box of a roll of kitchen film).

Torsional stiffness is most important in a car chassis since it gives a chance for the designer's suspension system to work correctly. It is no good designing a clever suspension which theoretically, say, keeps the wheel at right angles to the road, if the chassis it is attached to flexes so much as to render useless the skill lavished on the suspension. Similarly the suspension will only work if the axle on which the wheel turns is properly controlled by it.

Many suspensions consist basically of our quadrilateral of links, often 'triangulated' by the telescopic spring unit. These parts normally give good rigidity because they are loaded in tension or compression and are thus efficiently used. The part which fixes the wheel spindle to the suspension links is the suspension upright which is loaded in torsion and bending. From the same reasoning as used above, many suspension uprights are made of a broad tube, flattened at one end to connect to the links and at the other to hold the wheel bearings.

Again, if a tube is good in torsion and this is how a chassis must be, why not a chassis which is one big tube? Hence the idea of a 'monocoque' chassis.

The principle of a tube, used in torsion, or indeed in bending, can be thought of as an attempt to get the material as far away from the load as possible. A certain amount of material spread thinly (forming a tube) is often stiffer – and sometimes stronger – than the same amount of material bunched together (forming a rod).

A monocoque chassis carries this idea to a logical conclusion by putting the chassis structure material around the outside, where it does most work. As with so many things, Nature got there first and uses the principle in eggshells and in the bodies of certain insects. Also plenty of plants have tubular stems giving strength and at the same time reducing the strength needed by having less weight. Another example of Nature having produced a very sophisticated structure is found on examination of the skeleton of a bird. The bones are hollow, with internal bracing at the points of highest load, and are very intricately developed.

Stirling Moss, in his book *My Cars, my Career* (Guild Publishing), said about chassis torsional stiffness: 'What was nice was to be able to see the wheels go up and down with nothing else flexing'. Writing of the G-type ERA he said its chassis had a torsional stiffness '... which would not be bettered for ten years, until the arrival of the Lotus 25.' On the other hand he says that the Cooper Alta Special of the following year '...demonstrated the torsional rigidity of a wet bus ticket. It flexed and whipped and vibrated under braking and ...was extremely twitchy if I tried to corner at all quickly.' Thus an expression of the driver's point of view.

Another technical term will be met in the text in connection with the cornering of racing cars. This is the way engineers measure accelerations. A corner is sometimes described as 'a .5 g turn' or 'a 1 g turn' and this measure refers to the ratio between the cornering acceleration and the acceleration due to gravity. To understand this it is necessary to appreciate that cornering, even at a steady 50 kph, involves acceleration. Imagine a car approaching an observer from his right and turning away from him in a right hand turn; then driving off to the observer's right again. When approaching, the car was going from right to left; after turning it was going from left to right: it has clearly accelerated in that direction, first to bring its left-right speed to zero, next to raise it to that at which it departs. Yet its own speed had remained constant throughout. Engineers use the term velocity to talk of movement relative to a fixed observer and use speed for movement as seen from the moving object. The car's velocity changes while cornering, while its speed may not. The change of velocity implies an acceleration and this can be measured in a simple way by considering how long the car takes to complete a U turn at steady speed. In doing this it has changed its velocity by an amount equal to twice its speed. Say it approached at 10 metres/sec and took 4 seconds to turn. Its velocity changed at a rate of 5 metres per second in each second; its average acceleration is thus 5 metres per second per second.

When an object falls under gravity it gains speed (unless slowed by the air) at a rate of about 10 metres per second. The car above therefore made a .5 g turn. This means to the driver that he experiences a sideways push from the car's seat equal to half his weight, and to the tyres that they are transmitting a shared sideways force equal to half the car's weight. Note the use of the lower case 'g'. Journalists, science fiction writers and even some technical writers often make the mistake of using 'G', which is wrong because in science it refers to something quite different.

Engineers, of course are of many different branches. They may be trained and practised in the mechanical, aeronautical, naval or other disciplines, and within those there are many specialities. Thus a mechanical engineer may know all about gearbox design but be (relatively) ignorant about the engine which drives it.

Engineering training, however, covers most disciplines to some degree and thus a specialist in one field can often be competent in others.

In particular, where that engineer by necessity takes an interest in another field he may well master it sufficiently for his needs. Further, when we are dealing with an engineer of Chapman's calibre he will quite often be outstanding in several fields related to his interests.

Chapman showed original thinking in many branches. Obviously in car chassis design, where he created numerous 'firsts', some of which changed motor racing permanently, but also in engines, aircraft, furniture, boats and coffins. Some of the ideas which illustrate his scope have been covered in a later chapter.

1 The Man

Colin Chapman was arguably the greatest contributor to the design of racing cars since the origins of the motor car itself. No other designer was as consistently successful in his pursuit of technical aims, nor has any other designer yet had such a profound effect on so many branches of the sport. These views are held by many motor racing enthusiasts and experts, not least by his own rivals.

His influence can be seen in the transformations that have taken place in many fields, from the modern Formula 1 car to a generation of road sports cars.

Many of the ideas which earned him his reputation are, in fact, 'just' new or different applications of well known principles. This does not however lessen their value or relevance as examples of genius, since the same is true for a high proportion of all new patents granted.

Chapman's originality showed itself in not being limited by any already defined rules. He often said that: 'The trouble with experts is they know what can't be done'. This is, of course, another way of expressing the perhaps better known phrase 'an inventor is someone who doesn't know that what he is trying to do is impossible'. The mentality behind this manifests itself in two ways. First in simply doing something that no-one has done before, second in finding a way to make something true that at present is not.

It was an example of this second form that brought Chapman into conflict with the rule makers. It is well documented that he pioneered the technique for separating the twin 'siamezed' inlet ports of the Austin Seven engine for 750 formula racing. Whereas others might have said 'what a pity the ports are siamezed' and left it as a regret, Chapman simply refused to accept these limitations.

That the modification was banned still raises a few eyebrows as it was well within the capabilities of most serious competitors. Indeed, others have extended it to the use of a four choke carburettor from an American V-eight. Chapman had only used a twin choke, but he did the groundwork which pointed others in the right direction.

It is perhaps unfortunate that Chapman was so perceptive and analytical in his approach to racing. More than once he created problems for himself in trying to find ways of using the rules to his advantage. It is true, however, that each time he did, he could be seen as showing the way for others to follow, rather than finding a way of cheating.

With creativity of this type, technical skills only come into play when the final solution is at hand. The 'invention' itself is entirely conceptual, an overview which enabled Chapman to ask the right questions.

Originality shows, therefore, in the skill of asking the right questions, rather than being held back by a fogged view of what the real problem is. Once problems can be stated in a way that attacks their heart, rather than addressing obvious but less fundamental areas, then the solutions arrived at will inevitably be original. What is more, a problem thus stated can often be taken over by a person who is technically skilled but has not the creative skills.

An unfettered mind is free to see the nub of a difficulty, and also to be on the look-out for clues in any field which will help resolve a problem.

A sort of creative 'suction' will therefore have been set up so that we have a third way in which creativity can show itself – in the ready acceptance of the idea of another. This ability is particularly

Lotus' second single seat car and last front engined one was the Mark 16. A Chapman design with a Frank Costin body shape, the car came after both of them had worked on the Vanwall. In many ways the same precepts were applied. The car was referred to as the 'Baby Vanwall' and when it first came out many were amazed at its compactness, saying: 'how do they get it so small?'. Only a few years later the car seemed quite bulky in comparison with the 21.

powerful when the suction means that greater potential can be seen in the concept than was seen by its originator; perhaps because he or she was solving a quite different problem.

To make a quick round-up of some of the ideas that Chapman conceived or exploited, the desiamezing already described would be included, as would the strut suspension and the inflatable tonneau cover for the Lotus sports cars for the Le Mans 24 hour race; also, of course, the (in)famous double chassis design for Formula 1 which caused the biggest conflict of Chapman's career and which will be described in detail in a later chapter. Also worthy of mention is his exploitation of the fixed length drive shaft (which Eric Broadley pioneered on his Lola cars).

From the above it sounds as though Chapman's genius boiled down to a handful of ideas and this perhaps highlights the true nature of such a man: that throughout his life he showed a determination to achieve his aims through the application of sound principles, together with an ability to produce original concepts when nothing available met the needs.

Genius has been described as an infinite capacity for taking pains and that this is accurate is fairly evident in the creation of a sculpture, painting, or piece of music. In this type of creation it is obvious that what has been at work is an absolute determination to produce something the artist has in mind without compromise. The result is then a finished work which is valued and appreciated by the world at large. What the public does not see is the process of trial and error that so often takes place to achieve the vision.

Engineering is sometimes like this. The engineer's aim will be the ultimate simplification of some design feature and it is therefore hard to see in the finished result that anything creative has been achieved. The way an engineer demonstrates the 'infinite pains' is in his refusal to accept compromise. In effect, he never says 'that'll do' unless the aim is realised. Given such an aim he will try all possible approaches to the problem and each new approach will either be nearer the aim or will add a small step towards understanding it.

The process of design, following the conception of a new idea, flows something like this: The idea is formed and its existence poses the question 'how can I make it work?' Thus this is the first aim.

When it works the next aim is to make it work better. The aim here will be built upon the observation of the way the idea meets the designer's first aim, and the refinement of the aims through a growing understanding of what aims the creation might fulfil. In practice this phase of development may be the longest and may be a lengthy reiteration of study-modify-study, until the performance aims are met.

A challenging feat of escapeology is the best way to describe how an engineer wrestles with a particularly difficult set of compromises. He chains himself up with the conditions the idea must meet to fulfil its objective of 'working', and each condition will involve some conflict with the others, or with some external limitation.

For example, a theoretical solution to the requirement to make something work may be impossible to manufacture or assemble. The engineer must then 'juggle' other factors to solve this limitation. Such reiteraiton is a normal part of any engineering design work, but it is certainly true that the best designs—those that satisfy most objectives or that compromise them least – are achieved by those engineers with the most determination to wriggle out of the constraints. Chapman had this ability in full measure.

A prime need of any engineer is the understanding of his craft. It is also true that many times during a career, problems are solved not by a technological breakthrough, but by an ability to see through a confusion of factors to the simple solution hidden within.

Chapman often spent time with a less experienced designer working under him, leading the 'disciple' through the study of needs and constraints. This process went on until the designer would come to the key needed to solve the puzzle. This point reached, Chapman would produce another of his epithets – 'You see, it designs itself!' and then walk away.

This saying will strike a chord with many engineers and its truth will be recognised as a fundamental of the 'good' design method. Basically if what will not meet the objectives has been fully defined, the designer will fall almost inevitably upon that which will do so – thus the thing designs itself.

Many creative people turn their interests to things technical and mechanical. They will often come up with original ideas not seen already by 'experts' in the field (because they know what can't be done?) but be unable to put them into practice because they have not the detailed knowledge necessary. Sometimes the ideas will be impractical because they are based on a false premise and represent something that simply cannot function – such as the innumerable perpetual motion machines. The broader a person's skills and background of knowledge the more chance there is that his ideas will be practical and valuable.

Colin Chapman, through his extraordinary understanding and breadth of technical knowledge, was able to produce ideas that were intrinsically valuable and at the same time either practical or within his skills to make practical.

Being able to come up with original solutions when faced with a problem is a capacity held by many competent engineers and by most of those who, by their particular mix of abilities, practise engineering design as opposed to other vital roles where creativity is not a key element. This ability is not the stuff of genius. To become that it must be supported by an ability to recognise which are the problems which need solving. In the case of Chapman, he was brilliant at *revealing* the question at the crux of a problem.

Coming back again to the simple example of the de-siamezed ports as a model to explain this property: a less forward thinker, looking at the inlet system of the Austin Seven might have seen the problem as 'how to get maximum power from an engine like that?'. Chapman however must, in effect, if not consciously, have asked 'how do you design an inlet system for maximum power?'. The answer to the first question leads to a meticulous study and development programme resulting in some useful improvements. The answer to Chapman's question however is simply 'by having four separate inlet ports to maximize mixture flow', which then leads to the comparatively simple technical question 'how do I convert siamezed ports into separate ones?' and we know the answer to that.

It is not intended that this book eulogise Chapman's creativity; it is intended to be a reasonably balanced look at the man and his technical achievements. Chapman, of course, was not perfect; he had his weak points as an engineer and as a person. As an employer he was one of the most rewarding people to work for. He had the ability to make one really pleased to be part of his team and to share the excitement of the creation of sometimes sensationally new products. He could be, though, the most infuriating of bosses. This came about most when he was juggling several priorities and controlling a number of engineers on different jobs all at once.

There were times when your project was at the centre of his attention – then you felt very much involved and on very close terms with him. He would chat freely about things he was involved in or just on general topics, exchanging jokes and so on. Another day, maybe the next, when he was preoccupied with another job, he might be most distant – even rude; maybe pass you in the factory yard without giving any sign of noticing your presence. He was the man who could turn out at midnight to visit a team he knew to be working, bringing food and drink; just as easily he could turn on the stiffest formality to administer a reprimand he believed necessary. It has to be added that at such times he was not ready to listen to anyone trying to put another interpretation on the events which had incurred his wrath. On one occasion he sacked a young mechanic on the spot after he drove a Team Lotus Lotus Cortina into a wall, not waiting or wanting to hear that the car had been left in the yard with the brakes disabled and the key in!

On one occasion during the early days of the collaboration with Ford on the Lotus Cortina he was particularly stretched. He had impressed on his team the professionalism of Ford and was emphatic that the best possible image should be projected towards Ford personnel. Early one afternoon one of the Ford bigwigs arrived in a chauffeur driven car, asking to see him, presumably by appointment. Chapman was in his office, holding some meeting and would not answer his secretary's call. She then asked a member of the design team to look after the visitor until Chapman became free. Accordingly, with many apologies, the visitor was given a tour of the Elan build shop and the racing car build area in Lotus Components. Still no Chapman. A demonstration run was arranged for him in an Elan and in a Lotus Seven. During each of these diversions the embarrassed host went to remind Chapman, on the third occasion, braving his wrath and against the pleas of the secretary, knocking on his door. This was met by Chapman appearing and saying 'Look, if I don't answer my buzzer it means I don't want to know – doesn't it?' Finally after about two hours when the patience of both the visitor and the enforced host were wearing thin, they were standing in the factory yard when Chapman came up at the trot and said a friendly hello to the visitor and '... Sorry to have kept you, no-one told me you were here!'.

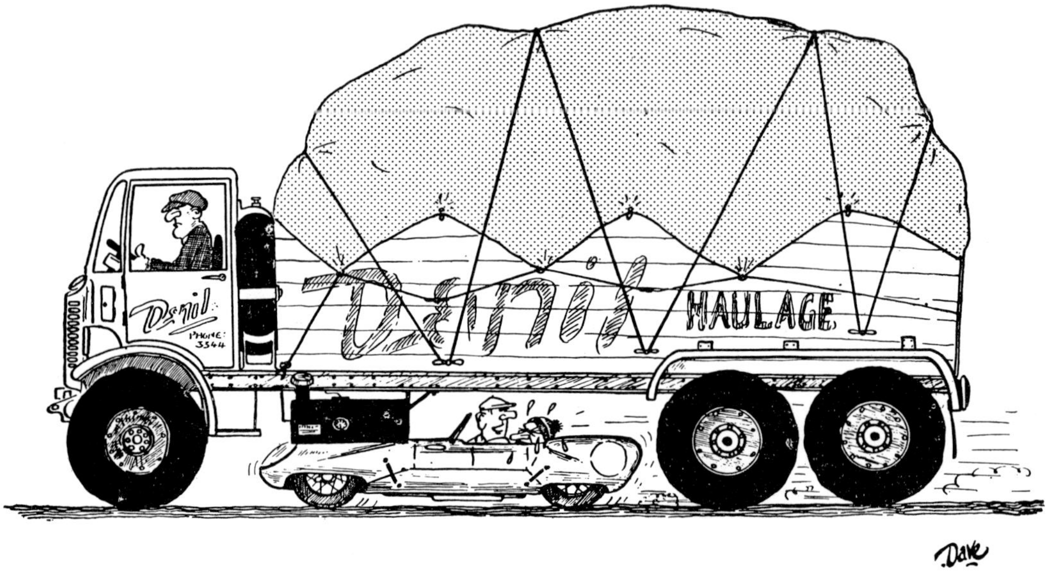

'WELL WHADYA KNOW – INDEPENDENT ALL ROUND!'

Lotus cars have made a special impression ever since the earliest days. This cartoon highlights the lowness of build.

2 *The Beginning*

MANY CHILDREN SHOW early signs of outstanding ability, but only a few go on to become such high achievers that they earn international renown in their field. Much has been done by different workers in different areas of study to try to pin down just what it is that transfers a person from the category of having a very high potential to that of genuine high achiever. This work has uncovered a number of elements which help towards this result. Parents' education, attitudes and economic situation, schools, opportunities for developing aptitudes; all play their part along with thousands of other factors.

However, these are the external influences; the internal influences – the nature of the person concerned – are clearly at the centre of things since it is these which allow him or her to take advantage of the outside influences that touch them.

In particular, what makes the difference between a clever person (even an outstandingly clever one) and someone who makes a real impact on the world is more to do with determination than the more obvious manifestations of ability. An iron will, an unswerving aim, a burning passion – these are the terms used in describing people who have made a notable contribution in their field. Single minded pursuit of a 'better' result from one's activities and not being satisfied with less than complete success separates the great from the competent.

Many learned works describe the psychology of genius and its development. For the purposes of this book it is sufficient to make a few layman's observations.

As the young man develops from the child, a focussing of ideas can be observed; a closer grouping of subjects of interest and a more objective approach to the solution of problems making use of any innate abilities.

This process is self sustaining since the creative young person, finding an aptitude in some field of endeavour experiences an intensification of interest. This in turn creates more opportunities for exercising the abilities, reinforcing the interest, skill and satisfaction.

Anthony Colin Bruce Chapman was the only son of Stanley and Mary Chapman, owners of the Orange Tree public house at Richmond in the westerly suburbs of London, some ten miles from the heart of the capital. He was born on the nineteenth of May 1928, his family moving to take over the Railway Hotel at Hornsey, north London while he was still very young.

Both of the pubs his parents owned are still in existence today, still bearing their original names.

It is quite evident that Colin Chapman shone as a child. This can be gleaned from a number of small clues which have come to light. From a school report, shown in Jabby Crobac's excellent biography *Colin Chapman; The Man and his Cars*, we can see that at ten years of age he was coming top of his class with an average mark over 80%, peaking at 98%. The childish sketch showing how to make a figure from a cork is of high quality for a boy under the age of six. Trace through the steps outlined (the sketch reads a little easier if you think in a cockney accent!). The writer takes us through the sequence quite clearly:

'Get a cork, a nail and a piece ('pes') of wire.
Wind the wire round the nail.
Get some beads and tie them on the wire.
Get a stick and put it in the back of the cork.
And then your little man is done'.

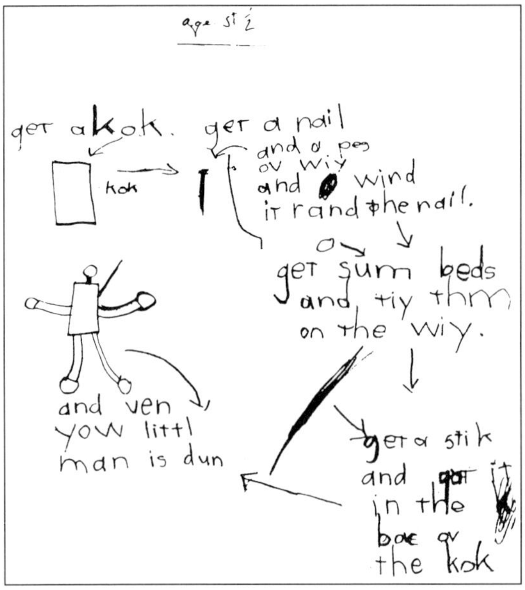

A drawing made by Colin Chapman when he was five and a half-years old.

These instructions and sketches were made by a child of five and a half and we must remember the context is not known. He was no doubt reproducing something already described by the teacher but we can perhaps make a few tentative observations from it. The writer was trying to explain a physical process. He therefore had to choose appropriate illustrations which would show what he wanted. Where the drawing did not show enough he added words to supplement it.

The information is put in the correct logical sequence for the task to be carried out efficiently, and the drawings are put in positions to support this. The drawing shows a clear understanding of the physical objects involved and the method to be used. A good idea of scale is present and—most revealing—the whole is in the form of a flow chart.

It is interesting to note that this (allowing for the differences due to the young age) was exactly the type of information Chapman was to use to convey his ideas to his designers for many years and over many different products. Sufficient information is given to convey the ideas without wasting effort on detail which the intended user is assumed to have the skill to be able to supply.

Later, as a schoolboy, Chapman stood out from his fellows. He was a friendly boy, full of fun, and it was always he who was the leader or the originator of escapades.

Doug Springham, a contemporary of Chapman at the Stationers' Company School, Hornsey and a member of the Air Scout Patrol which Chapman came to lead, has many fond memories of this period.

He remembers Chapman as always laughing and ready to tease, but having powers of understanding way beyond all the others, with no trace of immodesty. Chapman's abilities were further enhanced in the eyes of his contemporaries by his single-minded approach to everything he tackled. This is not to say that he needed a particularly intense level of concentration in order to master something, rather that he naturally put all his considerable powers to work for the result he sought, and consequently did things rather better than most.

Being the son of pub owners, obliged to commit long hours to their work at times of the day when other parents can free themselves to be with their children, Chapman was left very much to his own devices. Having no brothers or sisters added further to the time he had alone. Some children would have used the freedom to get into mischief or become bored, but Chapman's natural curiosity and powers of learning led him to explore, study and reflect. He discovered that most things of a technical nature could be worked out from an understanding of basic principles and his ability must have spurred him on to more learning.

Chapman made outstanding progress in the Scouts. In the scouting movement, Scouts measure their progress by the earning of badges. These tokens of the passing of tests of learning are of course sewn on the sleeves of the Scout's shirt. Chapman very quickly covered his shirt sleeves with the badges that were so hard for the others to win. He went on to obtain the coveted Bushman's Thong and received the top honour of being made a King's Scout.

For the most part Chapman found everything easy at that time of his life. The earning of scouting badges he took in his stride, his friends didn't even realise he was preparing for them, it came so easily. Only once does Doug Springham remember Chapman working for a badge. This was for the Pathfinder's Badge for which he was to be tested on his knowledge of the locality where he lived. On a casual call Doug found Chapman poring over a map of London – he had chosen this as his area, practically the same test as would be taken by a driver applying for a London taxi licence.

Chapman later joined the Air Scouts as a natural combination of his pleasure in scouting and his aeronautical interests. Here again he excelled, quickly becoming a patrol leader.

The members of his patrol soon came to appreciate the benefits of his leadership when they started to win many of the inter-patrol contests that enlivened their meetings. An outstanding example of this was the Trek cart race. This was a sort of small scale version of the famous race in the Royal Tournament in London where teams compete to convey their field guns over obstacles using manpower alone. The Tournament is an unusual event in which teams of highly trained and rehearsed Marines race each other in an arena. The guns are dismantled, swung on derricks over a mock river, reassembled and fired in only a few minutes.

The Trek cart race followed much the same pattern; the boys had to get their carts over a number of obstacles using only poles and ropes. The carts were heavy and had to be dismantled to get them over the obstacles by swinging them from tripods made of poles.

Chapman did not allow the team to muddle through with improvisation as so many would. Once he had understood all the steps needed he took time to work out the choreography of the whole exercise from beginning to end. Each team member was then assigned a precise series of tasks, carefully thought out to mesh in with those of the others. The team was rehearsed under Chapman's close scrutiny until they worked perfectly together, developing the moves and their interactions by trial and error. Each step was worked at to perfect it, and was timed to ensure a proper distribution of work in the team. As might be expected with such a detailed approach, Chapman's team came out easy winners.

The boys in Chapman's patrol, which he christened the Gannets through his love of the bird's

A line-up of contrasts. A Lotus 18 is seen alongside a Mk III and a Mk IV.

Colin Chapman's first taste of motor sport was in trials. Once he 'discovered' racing, however, he ceased building trials specials. He did not abandon the hills altogether, though, and is seen driving a Cannon in a trial in 1961.

qualities, were inspired by his enthusiasm and thorough preparation; and seeing the results they achieved, buckled to with a will lacking in other patrols. In the words of Doug Springham, 'there was no competition in anything – whatever we did we were on the winning side, and loved it'.

One of the interests Chapman had from about the age of twelve was making model aeroplanes.

Here again he was remarkable in the eyes of his friends. Not only was he a prolific producer, but his models were made to a level of perfection that they could not match. Whereas most of them, with boyish impatience, would be satisfied with models hastily painted in cellulose dope applied directly over the wood or paper surface, Chapman's models were often carefully filled and sanded during construction to produce a perfect finish. This was of course a normal procedure for the older modeller – but it is unusual in a boy of his age with no-one as tutor.

Thanks to his combination of curiosity and ready understanding he was able to see possibilities in things that his friends would not. He collected interesting oddments from scrapyards, behind factories and so on, always looking for something new or different. Thus when he found some scrap perspex in his wanderings he experimented to see to what use it could be put. His models were the first among his circle to sport perspex cockpit covers.

Chapman's home was a magnet for his friends who were drawn there like iron filings. But he had no arrogance for all his outstanding abilities. Even at that age he was a natural leader who could set examples and pass on ideas, instilling enthusiasm at the same time, rather than appearing clever or bossy while doing it.

In an idle moment the boys of Chapman's patrol would drop round just to see what he was up to, knowing that it would always be novel and interesting. On one of these occasions Doug got round to asking Chapman about the drawing that hung over his bench. It was a technical drawing showing the details of a machine pistol. Apparently Chapman had acquired enough draughting skill even then, without formal training, to have produced the drawing which, he said, he had submitted to the Ministry of War (today's Ministry of Defence), since the country was at the time in the midst of the Second World War.

One day, when Doug was in a group in Chapman's cluttered room, something happened which shines in his memory as a symbol of how life should be approached. The boys had been discussing some new modelling project around the workbench which was, as usual, covered with tools, tins of paint and general modelling debris. When the talk had reached the point where action was needed, Chapman said 'Right, let's get started', and with a sweep of his arm cleared the bench. To Doug, at least, the message was emphatic: clear the workplace, clear the mind, setting aside other problems, and attack the new work in an uncluttered way. Chapman's ability to do this metaphorically, if not physically, was characteristic of him in later years as he concentrated single-mindedly on the problem in hand, not letting the other 'rubbish' in his mind detract from it.

Whatever he was doing, his companions found Chapman a pleasure to be with, always laughing and the centre of attention. In the course of their scouting training, one of the challenges they were given sometimes was to hike from one point to another at night. Chapman enjoyed leading these but on one particular occasion his natural quickness let him leave his team in the lurch, which later made him the cause of much merriment within the group. On this night they became somewhat disorientated in a mist that had come down and found themselves in a churchyard – very frightening for a bunch of boys. Out of the darkness they heard a large dog barking fiercely and charging towards them. Chapman's reactions and athletic build allowed him to escape by leaping the barbed wire fence, a feature of war-time Britain as a hindrance to possible invaders, while the others were still wondering what to do. To the relief of the others and to the embarrassment of Chapman, the dog was friendly and his precipitate flight unnecessary.

On one training evening it fell to Chapman to teach his patrol about the internal combustion engine. Other patrol leaders would run this session using just a blackboard and the various diagrams provided. This was not good enough for Chapman however. From some scrap yard or other and with the help for once of his father for transport, he

Originating in 1953, the Mk VI became the standard Club racer of the mid 1950s. This example is seen at Bottom Bend, Brands Hatch and its left front wheel is off the ground.

brought an almost complete engine to strip down to help explain to his group. In this typically thorough way he explained the crankshaft and pistons, the valve gear and other parts.

Even while apparently serving his patrol so well he could not resist a joke. Taking advantage of their lack of knowledge and the fact that they would readily do what he asked, he suggested checking if the magneto worked by them all linking hands and holding the terminals while he spun the shaft. Once they had got over the surprise of a high voltage shock they all had a good laugh – but none would try it again.

The ability to draw people into his enthusiasms and to fire them to greater efforts in the pursuit of success was with him all his life, and it is interesting to observe at what an early age it showed itself. Histories of Chapman and of Lotus are full of accounts of how some superhuman effort was put in to finish a job in time or to come up with a solution to a knotty problem, and of his ability to talk reluctant collaborators, sponsors or drivers round to his point of view.

Chapman's first motor manufacturing activity was in about 1947 when, following the crash of his second hand car 'business', he tackled the adaptation of a car to his own liking. He had been trading with a fellow student, Colin Dare, during their studies at University College, London. Their little affair became impossible when the already meagre petrol ration was cancelled. Chapman then decided that one of the cars on hand, a 1930 Austin Seven, would be the basis for a trials special. This car, built in a lock-up garage in the spare moments of his engineering course with the help of neighbours and friends (including his future wife Hazel Williams) became the Mk I Lotus.

Following his graduation with a B.Sc. in 1948, Chapman set about filling out his new found knowledge through the study of technical journals and learned papers. He joined the 750 Motor Club (which had been formed originally in 1939 to further the interests of impecunious motoring enthusiasts), where he learned of many problems

The Lotus Mk VI was the car which first made the name famous on the track. H. P. Deschamps is seen winning one of his two first places earned at Snetterton, 15th April 1956, in his Climax-engined example.

(and solutions) encountered by special builders. Against this background, Chapman was formulating ideas for another car as a step on from the first, using his growing fund of knowledge. The car was entered in some of the popular trials organised by the many motor clubs around London in those post-war years; trials that took place on muddy, tree covered hillsides.

The new trials car was Ford powered and was built despite Chapman's having been called up by the RAF for his National Service. In the Air Force he was soon moved to Market Drayton in Shropshire and it is said that on Hazel's dates with him during this period she had more job lists and instructions than romance.

It was this car that benefitted from one of the first known 'Chapmanisms' in car design. He had worked out that the car would need an axle ratio of around 4.5:1. However, the only Austin Seven ratios were 5.125:1 and 4.89:1. These used an 8 tooth pinion for the first with a 41 tooth crown wheel and a 9 to 44 tooth pairing for the second. It did not escape Chapman that taking the 9 tooth pinion from the higher ratio to run with the 41 tooth gear from the lower one would give a ratio as wanted of 4.55:1.

Every engineer (including Chapman) knows that you 'can't' mix gears that were not designed to go together. The tooth shaping would be quite incompatible and the combination would be a sort of technical sacrilege, to say nothing of the risk of failure. However, Chapman also knew that if he was not concerned about noise produced by the axle, nor about it achieving the maximum possible life, that it would work. All that was required was to force the gears to wear each other to a less mutually conflicting form. He acquired two complete axles to provide the parts needed and built up his hybrid. He installed it in the car and filled it with metal polish in place of oil so that a few miles of running would go some way towards the needed re-shaping of the teeth. The ball bearings were ruined, but as he had two axles he had a second set with which to replace them when he stripped it down later. Refilling with oil then gave him the axle he wanted.

This type of pragmatic approach to problem

Just as the 6 was the club car of the 1950s the Seven took a similar role in the 1960s, continuing to the present day.

solving, while horrific to the purist, was a lifelong characteristic. It was in some measure his willingness to expand on classic theory through any means available to him, and partly his carefree attitude to convention and what other people might think, which helped Chapman avoid getting bogged down with known limitations.

A completely different aspect of Chapman's abilities showed itself when he came to replace the Ford Eight engine of the Mk II with a larger one from the Ten model. It was the middle of 1949 and (as was usual) at a time when Chapman's needs were well ahead of his purchasing power (this was to be the normal situation for the Lotus company as a whole for many years to come).

Having scoured the breakers' yards, Chapman had found a suitable car which had been burned out, for sale for £30. He had discovered that, bad though the car looked, the engine had been replaced only 1000 miles before. So the source of the engine was established, but not the source of the £30. It turned out that there was a dealer looking for scrap cars for restoration for whom the most important parts were the registration documents. Chapman offered him 'his' Ford Ten less engine and gearbox for £35, which he accepted. Now there were two people to whom something was promised but no way to bridge from one to the other.

Chapman was able to persuade another breaker whom he knew to help. This man bought the complete wrecked car for the £30 asked and allowed Chapman to remove the engine and gearbox. The first dealer could then collect the remains along with the documents and pay the promised £35. The friendly breaker kept the £5 difference for his trouble and Chapman had a free almost new engine with gearbox. To cap it all, he was able to sell his now unwanted 'Eight' engine to his friendly dealer for that £5 and thus actually gained £5 in getting a better engine. We can see the application of the ideals of practical engineering to the early Lotus chassis designs, even when they were still built on a ready made base. The tubular stiffening to the Austin Seven chassis for the Lotus Mk 3 can be seen as an example of an unconventional combination. A designer bound by the 'rules' would try to stiffen a channel section chassis by additional channel cross-members or other bracing. Chapman, however, knowing the advantages of tubular structures did not hesitate to add a tubular part to a channel chassis.

During the construction of the Mk III every part was weighed before it was passed for installation. Any part which did not come to what Chapman judged to be an acceptable weight was worked upon until it did, or was replaced with a more suitable alternative. The body of this car weighed only 30kg (65 lb), some 30% less than those of most similar cars of the type.

It was the Mk III Lotus of 1951 which benefitted from the de-siamezed inlet ports previously discussed. With its engine improved in this way, the car romped away with its first race to such good effect that only the second place man was on the same lap at the end of the ten laps. A few weeks later on 2 June was the annual meeting organised by the Eight Clubs at Silverstone, in which was an important 750 Formula race over five laps. During practice the car suffered a failed big end bearing. Rather than not start, Chapman removed the piston and connecting rod and ran the car on three cylinders. The power output still proved enough to put Chapman a long way in the lead when on lap four the vibration caused by the imbalance finally broke the maltreated crankshaft.

This third Lotus was the one that truly turned Chapman's interest to motor racing in particular, rather than motor sport in general. Despite its success the car was not made solely for racing but was also used on the road, even for a holiday for two.

Although Chapman now had a growing commitment to circuit racing, the next cars he built (in 1952, the year in which the Lotus Engineering company was formed) reverted to the trials field. This was due in part to commercial factors – that is to say that Chapman wasted no time in accepting an order received for two trials cars. The Mk IV had further small design innovations that show the Chapman touch. A need with trials cars is a very large relative suspension movement between front and rear, to accommodate the uneven terrain. Chapman realised that this articulation could be further enhanced if the front axle was allowed to pivot freely at its mid point as was normal on agricultural tractors. Normally the transverse spring was fixed at its centre and the axle movement was accommodated by the bending of the spring. On a tractor, there being no suspension, the axle is mounted on a pivot, so Chapman mounted the spring on a pivot, thereby allowing more and freer articulation. The device which permitted this became known as the 'jelly joint'.

Another small example of ingenious attention to detail which characterised Chapman's designs all his life was the adoption of hinged attachments for the front cycle-type mudguards. This trick allowed the effective width of the car to be reduced for narrow competition sections between trees.

3 *The Skills Develop*

COUNTLESS EXAMPLES can be seen in Chapman's work of how his aims have been raised as his understanding of the goal has grown.

To show this process a little more clearly, however, we need a specific example.

Looking at the front suspension used on many Lotus models from the Eleven to the 16 – known as the Baby Vanwall because of its similarity to the Vanwall itself (both bodies having been designed by Frank Costin) – it can be seen that Chapman opted for combining the top wishbone with the stabiliser bar. This was to satisfy the aim of simplicity and hence, lightness.

To digress for a moment, this layout is identical in concept to the lower part of Ford's McPherson strut used on many of their saloons. It seems highly probable that Chapman adapted from this system.

Once the aim of simplicity had been met, the suspension design served well until the early 1960s when, due in part to the progress in tyre design and the increase in wheel rim widths associated with it, the shortcomings of this suspension design became a limiting factor. These shortcomings were that it offered insufficient stiffness for the wheel location (which becomes more critical with wider tyres) and that fine tuning of the suspension (also becoming more important) was more difficult for two reasons. The first problem was that if the stabiliser bar was changed to alter the roll-stiffness, it could affect the suspension in other ways and tended to make the finding of correct settings more difficult, and the second was that the task of changing a bar, which was also a location member, took longer. There was also the additional problem that the castor angle was dependent upon the manufacturing tolerance of the bar and not easily made adjustable.

These considerations generated some new aims for front suspension design – or in fact a re-prioritisation of the original ones.

It is no doubt true that stiffness was among these but was satisfied at the level then required in the earlier scheme. Thus we come to the suspension seen on the rear engine Lotus 18. This design differed in having a full top wishbone and separate stabiliser bar. A gain was made in location stiffness and in ease of suspension tuning at the expense of simplicity and lightness, which now would come a little lower on the priorities list for the net benefit of the car's performance.

Up to this time the wishbone pivots had always consisted of rubber bushes which satisfied the needs of the time. A year or so later we see (provoked by the continuing trends in tyre and wheel progress) the replacing of these bushes by metallic ball joints so as to give a matching improvement in location stiffness. From that point on all Lotus racing cars had 'non-flexible' suspension bearings.

This balancing act between aims and necessity mirrors the evolution of most other makers' cars. It does not constitute anything specially 'Chapman' but illustrates the progressive shifting of priorities in the different aims of a functional design in response to development.

Where Chapman scored over many others was in his thorough understanding of the overall needs of a racing car which enabled him to anticipate changing priorities and so switch his attention to a new 'number one' sooner. It was this ability, perhaps, more than creative flashes which kept him ahead.

As with all creators, Chapman suffered from his failures. Some of these were brought about by his over optimism concerning a basic design and some

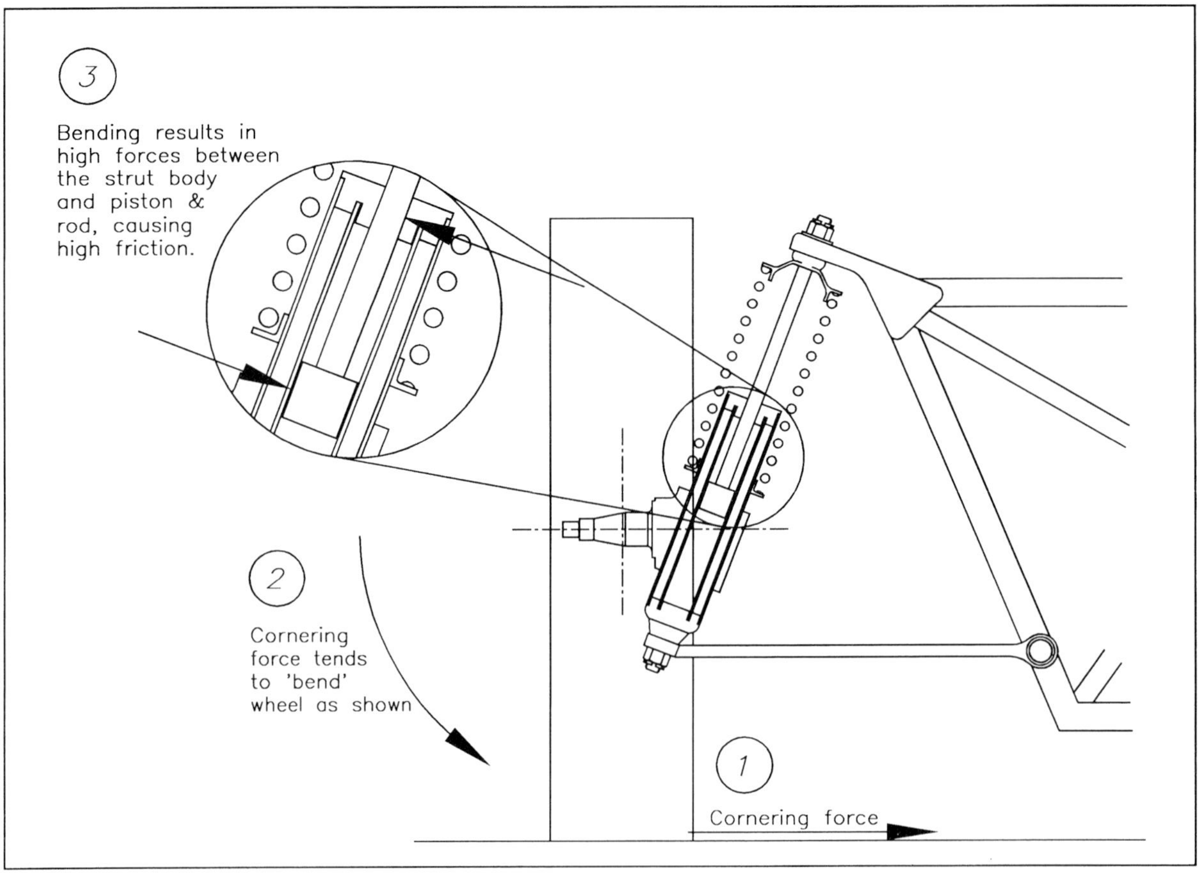

The Mk 17 Front suspension. On the 17 Chapman tried too hard to achieve a low body line while using the strut suspension in which he had great confidence. In shortening the strut to this extent he created a system that could not work freely. Free operation is absolutely critical for a suspension system if it is to give safe and predictable handling. This, in turn, drivers need if they hope to get the best out of the car.

by unwarranted enthusiasm in the reactions of others.

The Chapman Strut suspension must be rated as a considerable overall success but included at least one failure. In particular, the Lotus 17 used strut suspension at the front in a new configuration. This time, in order to maintain the minimum height of the front of the body, the struts were greatly shortened. This brought the bearings within the damper so close together that there was a tendency to jam the sliding of the damper.

A direct result of this was that when subjected to the bending loads generated in cornering, just when the suspension needs to work at its best, the strut became more like a rigid member than a sprung system. Thus any road bump tended to cause the wheel to skip and to lose its grip.

This was a contributing factor to the failure of this car. The 17 was designed at the height of the rivalry between Chapman and Eric Broadley of Lola for supremacy in the 1100 cc sports-racing class. The Eleven had dominated so completely that it became practically a compulsory choice for those hoping to do well in the class. Eric Broadley, however had responded with the magnificent Lola 1100 which Chapman recognised as a truly outstanding front-engined design.

Broadley's new creation was lighter than the Eleven it was designed to challenge and was of great stiffness as well. It was a very 'pure' tubular chassis design and extracted the best from the materials used. Added to this was handling and road holding of the highest order, meaning that with the right

Sorting out The Seventeen

A DESCRIPTION OF TESTING AND DEVELOPMENT WORK—BY ALAN STACEY

Leaving the "pits" for another session to find the effect of the latest "mods."

Discussion period; left to right, Graham Hill, Don Badger of Dunlops, Alan Stacey, Colin Chapman and another Dunlop representative.

Almost neutral - steer characteristics at Becketts.

I WAS very pleased when David Phipps suggested that I should contribute a few words on the new Lotus Seventeen, if only to help reduce the endless questions from enthusiasts and prospective customers—" Is it quick, yet?" . . . " Will it beat the Lolas?" As it hasn't done the latter yet, although I think it is certainly capable of doing so, I must try to give an idea of the development work which has taken place on the car.

QUICKER

To start with let me just say that I believe it will be quicker than all its predecessors. To all the pessimists and members of the Lola fan club who suggest " It's too narrow," I would say "What is the fastest car round a corner?" —a '500 with a 3 ft 6 in track. The point is that, as cars are going increasingly quickly these days and reaching a higher state of development, there is little chance of an entirely new design being " right " straight off the drawing board.

The need for a certain amount of modification became apparent in practice for the Goodwood Easter Monday meeting. I managed to record 1 min 40.2 secs despite atrocious handling qualities, but decided it would be unwise to race the car and treated practice as a testing session, which allowed me to make a few suggestions to the powers-that-be.

The next meeting for which the Seventeen was entered was Aintree on 18 April, and although I was busily engaged preparing the new Fifteen I was assured that all my suggestions concerning the roadholding of the Seventeen had been carried out.

In practice I thought we really had got somewhere. It was wet, but I had little difficulty in recording fastest practice time. However, in a " free for all " practice session in the dry the increased cornering forces brought out the old oversteering tendencies. Nevertheless, I decided to race the car, and although I lost my pole position on the grid due to some last minute carburetter trouble, starting from the back row, I managed to take the lead within two laps, when second gear went. Still, the car had shown its potential.

SUSTAINED EFFORT

Practice for the Silverstone International Trophy meeting revealed the same interent trouble, and now I had got my Fifteen going quickly, I decided a sustained effort would have to be made on the Seventeen. After two days' practice, with Mike Costin working full time on the Seventeen, we discovered the cause of the disconcertingly sudden oversteer. Unfortunately, in the time available this could only be cured by adjustments giving a completely oversteering tendency. The race thus became a test again, but it did prove that the " twitch " had gone.

After Silverstone, due partly to the not undeserved adverse publicity and lack of improvements we decided to do another day's testing at Silverstone. An early start was made and Colin and Mike brought the car and two mechanics with them, complete with an assortment of springs and anti-roll bars.

The test began with a few laps warming up and making sure the faults encountered during the last race were still present, and sure enough the oversteer was still pronounced. The first step taken to overcome this was the resetting of the ride level. This was found to be very critical, and was definitely too high. Lowering it improved the handling immensely, but with a full load of fuel it was found that the rear of the car bottomed through Becketts.

To prevent this, stronger rear springs were fitted. This, we knew, would bring back the oversteer, but it is is a principle of such tests to try only one thing at a time. With the new springs fitted the car certainly oversteered, but it was more gentle and easier to control. We had two ways of getting back to an understeering tendency, either by fitting a stronger roll bar or by stiffening up the front springs. As the car was not rolling excessively, stronger front springs were fitted, and as a result, the car handled very well indeed. After 79 laps of the Grand Prix circuit throughout the day I got down to 1. 49.5 —this with an engine that was 300 rpm down on Derek Randall's standard-engined Seventeen. From this lap time it seems feasible to suggest that with a few more horses times of around 1. 48 would be possible.

Very fast until the gearbox packed up; the Seventeen at Aintree on 18 April.

In comparison with the Fifteen and Formula cars, the Seventeen now has much more neutral steering, that is, less understeer. I believe it is quicker into a corner, and that with the increase in roadholding gained with the strut type rear end, it was essential to incorporate something similar at the front to maintain stability through the corner at the increased speed. This season all the quick cars using independent rear suspension seem to suffer from understeer. I believe the Seventeen has the answer.

After a quick check all round on the engine in an attempt to regain the lost horsepower, we took the car to Crystal Palace for the Whit Monday meeting. Practice showed that the horses hadn't returned, but a time of 1. 4.0, under Keith Hall's 1100 cc lap record, proved that the roadholding was now very good. During practice private timing showed that the Seventeen was easily the fastest up to 1500 cc car, including Salvadori's Cooper Monaco and the Lolas, through the notorious Ramp Bend.

On Sunday we rounded up the horses ready for Monday and for the first time were feeling slightly confident. However, in the first 25 yards of unofficial practice the crankshaft broke. Who said there's no luck in motor racing!

Well, that's the story to date. We've proved that the Seventeen is the lightest 1100 yet built and also that its roadholding is on a par with the best. Doesn't that mean its the fastest?

SPORTS CAR & LOTUS OWNER JULY, 1959 159

Such was the notoriety of the handling of the Lotus 17 that this article appeared in *Sports Car & Lotus Owner* for July 1959, trying to show that significant improvements had been made.

driver this car was a match for the Eleven. Indeed, it was later shown to be capable of challenging the rear-engined Lotus 23! The 17 was designed to regain the lead in this class which was bread and butter to Lotus at the time. It was even smaller and lighter than the Lola and somewhat shorter. The handling did not come up to expectations due partly to the sticking of the struts and partly to its unfavourable proportions.

While the car was sensationally light (it weighed less than 750lb) it cannot be classed as a successful design.

Another complete car that never fulfilled its apparent potential was the Lotus 30 – and also its later version the 40. The 30 was a 'breakthrough design' in that it was an attempt to use some new chassis design principles for this class of car. It had a light glass fibre body and was in essence a two seat version of the Indianapolis car of the same year, sharing many of the mechanical elements including the Ford V8 engine. However, where the Indianapolis car had a full monocoque chassis like an oversized Type 25, the 30 had a backbone chassis in sheet steel based on the principles of the Elan whose design was roughly contemporary. It was lighter and lower than previous cars in its class and was expected to be capable of taking on all the opposition.

During the testing of the Type 30 at Snetterton came one of those mishaps which show just what stuff certain of the drivers are made of. Jim Clark was driving the car and was working at around lap record speed. After a number of laps Clark failed to reappear at the pits when expected. After a while the mechanics went off in the transporter (no-one else was using the circuit) to find out what had happened.

At that time the Snetterton Circuit had the longest straight in Britain, followed by the tightest hairpin. Jim was found at the entrance to the hairpin, sitting apparently calmly in the car at the end of a series of artistically criss-crossing curved skid marks. Clark's explanation of what had happened went something like this: 'Well, when I braked for the corner the car did not slow but went sideways instead. I thought that was not normal, so I released the brakes and tried again. I then found they really meant it so I spun to a halt'. The technical explanation was that a rear radius arm had broken! That this acceptance of danger was not an assumed attitude after he had had a lucky escape was clearly shown at another time when the 25 was being tested, as will be explained below.

Whenever the design of a complex device such as a motor car is carried out it will contain a mixture of success levels in its different parts. For example, on a particular design the front suspension may have reached a relatively high level of perfection while the clutch system leaves much to be desired. The ability to achieve sound designs for all the different functions within the vehicle is clearly vital; not only in terms of pure functionality, but in reliability, durability, ease of repair and resistance to abuse.

This means that even in these and other unsuccessful designs there were successes in detail terms. There were also lessons learned from both the successes and the failures. This, of course, is the learning process that leads from a primitive product to a sophisticated and developed one.

A self-made difficulty for the innovator is that he is not just trying to make an existing product better. He is trying to take design or concept leaps, not just development steps. This means that the products of an innovator are likely never to reach a pitch of reliability and consistency theoretically possible with the less adventurous. This will inevitably happen as one idea which has been developed fairly fully and is therefore near its peak of perfection is abandoned in favour of a new one with a higher potential. Worse – in a fast moving environment

Compare p.30 with the Volkswagen 'Beetle' jack. A car jack normally depends on a screw system so that there is no risk of slipping. In this design the arrangement of forces is such that a good grip is guaranteed using friction alone. Chapman inadvertently got very near this condition in his Mk 17 strut.

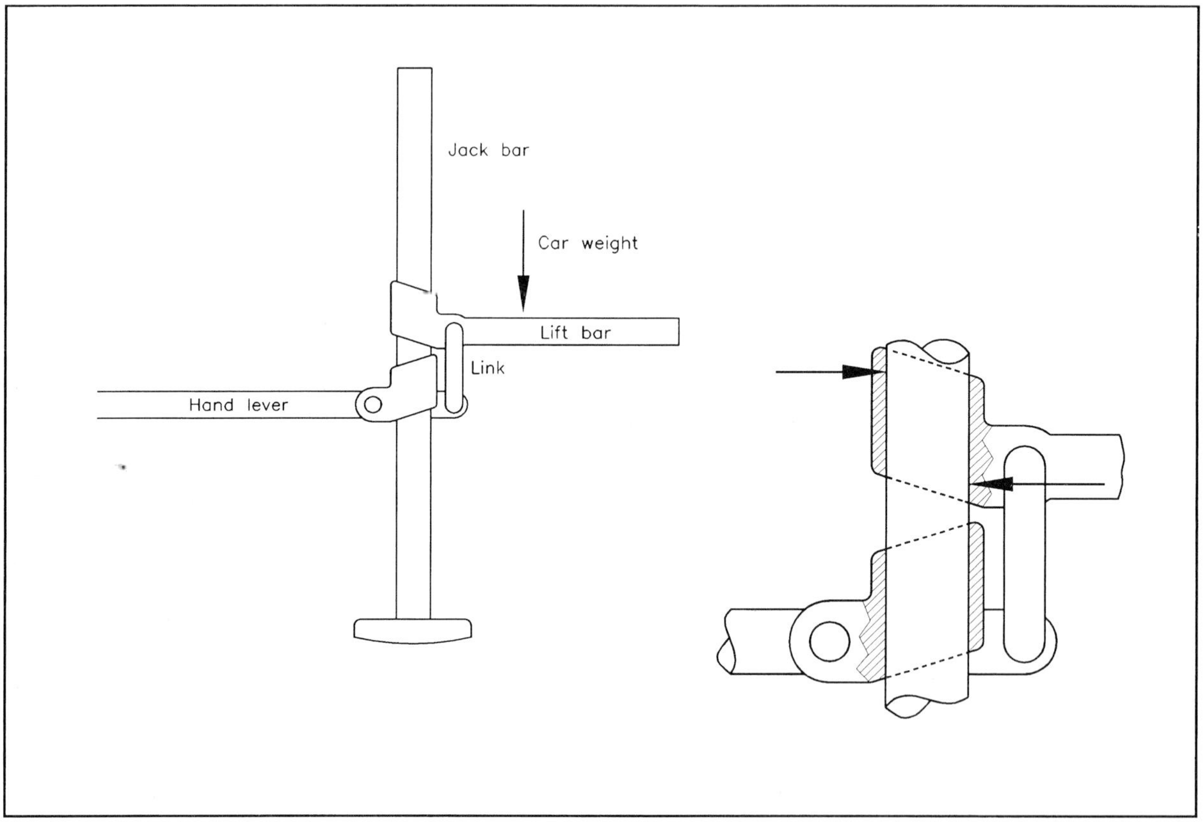

such as the motor racing world, few ideas actually achieve a full state of development.

The products of the major manufacturers (the better ones anyway) do rather better than this thanks to the large and expert development teams they are able to maintain as a result of their sales volumes. Small companies are always limited in their ability to develop their products.

One feature of the modern racing car that Chapman did not pioneer was the rear engine layout and it might be interesting to look a little more closely at this fact to try to determine the reasons why he was relatively late in adopting it.

The formative time which laid the foundation for the pre-rear engine days was right at the start of Chapman's car making. He followed the front engine principle simply because he was working from existing vehicles of this system. Clearly at that time, with the power-weight ratios at which his cars were operating, there were no handling or traction problems which could not be solved by careful attention to weight distribution and suspension design. Thus there was no problem present to cause a re-assessment of the layout. Cooper, on the other hand, opted for rear engine installation on the 500 cc formula cars as this afforded the simplest construction and drive configuration.

It may be that the two great protagonists on either side of the engine front/rear debate had taken their positions merely through the accident of which type of car they happened to start their careers on. It is intriguing to speculate, if the 500 cc formula had not existed and therefore constructors had not been faced with the problems of driving a four wheel vehicle from a motorcycle engine and gearbox, how long it would have taken racing car designers to switch away from the front engine layout almost universal until then (*pace* Auto-Union admirers!). Then again if there had not been a war and the consequent upheavals which affected the re-birth of motor racing that took place after it, would there ever have been a 500 cc formula to redirect design?

Once established with their respective choices of layout Chapman and Cooper followed their development paths. This as we know led to the point at which the drawbacks of the front engine position could no longer be offset by any tricks of the designer's trade. In any case most of the tricks available to the front engine brigade were also available to the other side.

Chapman's designs of front-engined cars progressed probably further than those of any other. There is an inexorable sequence over a lengthy period which can be seen in retrospect to be leading towards the choice of the rear engine position.

Starting with such early designs as the Eleven, the engine is placed 'comfortably' at the front with a conventional gearbox behind it, a short driveshaft and chassis-mounted differential. That the weight distribution tended to be too much towards the front is revealed by the fact that the driver sat well back and that the fuel tank, battery, spare wheel and any other movable object was also placed as near the tail as possible. The car used a de Dion axle.

The next evolutionary step was perhaps the 12 Formula 2 car. Here the de Dion initially fitted was later superseded by a fully independent design, partly to improve traction. The heavy parts of the car were brought more together and more rearwards, the gearbox being moved to the back to become a transaxle unit.

Following this design, Chapman tried even harder to match the almost effortless performance of the rear-engined car. In the Baby Vanwall Formula 2 car the engine was laid almost on its side in some versions, to lower its centre of gravity. This car had greater traction but lost some power due to the complex driveshaft layout. There was a driveshaft spinning at engine speed dodging past the driver from engine to gearbox.

Around 1958 it became fairly clear that the Cooper layout – and Cooper was about the only Formula 2 and Junior contender using it – had a better potential than that used by the majority. By then though, Chapman, forward-thinking as ever, was already working on his rear engine location design. Characteristically he was the first to make the change.

The first rear-engined Lotus was the 18, which ran in Formula Junior, Formula 2 and Formula 1 versions. The Junior version came out in 1959 and, in the hands of Jim Clark, regained domination of this class for Lotus.

In its Formula 2 guise, this car started to show up weaknesses in the Lotus transaxle. On earlier front-engined models, driving it through a somewhat whippy and power absorbing shaft, the unit had been developed to be quite reliable. Simply moving the engine to the back and coupling it directly to the gearbox input, increased the torque applied and at the same time increased the grip of the rear wheels. The combined effect amounted to such an overload on the gearbox that the internal shafts sometimes took a permanent twist.

This was the route taken by Chapman in working away from the front engine approach to the rear. Plainly, he felt some reluctance at doing what his arch rival Cooper had been doing for so long and in admitting that the firmly established layout was not ideal. Chapman's clear understanding of the fundamentals involved, in which he was not unusual, was allied to a single-mindedness of purpose, which was less common. This meant that he overcame this reluctance more readily than some others.

Although the trend just demonstrated is clear enough, there was another within it. This was the tendency towards a lower polar moment of inertia. This car designer's term, which is usually shortened to polar moment, describes the car's inertia in rotation. Taking any convenient object, a brick, a loaf of bread, or even a pencil in hand you can sense its polar moment. If you rotate the object about its length (spin the pencil like a drill) you can feel that it resists this movement only slightly. Turn it about its mid point like a propeller, though, and you can feel a higher resistance and a higher tendency to keep going. The difference is in the distribution of the material of the object. When the material is far from the centre of rotation it has a high polar moment and when the material is gathered closely round the centre it has a low one.

The reader may recognise the similarity between these ideas and the effect of distribution of material on stiffness already described. Indeed, moment of inertia is one of the pieces of information the designer needs in estimating stiffness.

In the case of the car, the gathering together of the heavier elements (driver, engine and axles) reduces its polar moment. This was happening as the weight was being moved rearwards mainly for other reasons.

But why might this lead to improvements in the car's performance? For a car to negotiate a corner it has to rotate about a vertical axis. For this to happen it must start its rotation as it enters the corner and stop on leaving. It is at these times of rotational acceleration and deceleration that the polar moment makes a difference. A high polar moment will mean that the car requires more from the tyres to make it start turning and again more when straightening up. A car with a low polar moment is therefore more responsive (but less stable) than one with a higher polar moment. Up to the point where the car becomes too unstable this extra responsiveness can be an advantage.

Recognise also that there is a similar notion of

The Lola 1100 was the only competitor able to beat the Lotus Eleven in sportscar racing in the late 1950s. Eric Broadley's car was lighter than the Eleven and the chassis design was more elegant and stiffer. This was the car that made Chapman design the 17, which, although even lighter than the Lola, had severe handling problems.

A Seven Series 1 in the course of assembly at Hornsey, circa 1958. This car has a Ford sidevalve engine of 1172 cc. The simple rod extension of the gearchange can be seen just above the carburettor. This type of extension was used regularly until the author fitted the remote gearchange adapter from the Triumph Herald.

moment of inertia in considering the effects of the tyres on the vehicle. The more 'spaced out' they are the greater is their power to command the vehicle's response.

Troubleshooting in Formula 1

Even before Chapman had changed over to rear engine layout, he had been given one of the greatest accolades imaginable for any racing car designer. He was hired as a consultant by no less than two of the main contestants in Formula 1. What is more, this happened *before he himself had even designed a Formula 1 car.*

This must surely be a situation unique in motor racing history. How could a designer, with no Formula One experience, find himself in such a position? In the mid 1950s, Chapman's cars dominated smaller capacity sportscar races and had accelerated suspension technology and understanding: the skills of extracting the best from the tyres in cornering. He was the master of suspension geometry, unsprung weight and the influence of the chassis. The main thing other experts saw was that the performance gains he made possible were not a question of hitting a lucky combination, but the result of a true understanding of the problem; he could therefore be expected to be able to use his skills to help them.

At this time Britain's two principal Formula 1 teams, Vanwall and BRM, were both striving to make their cars truly competitive. The Vanwall team had started by buying a Ferrari in 1950 and naming it the Thin Wall Special. It had developed the car through a number of phases which made it more and more special and less and less Ferrari. When the 2.5 litre formula started, the team created its own engine for the first time and named the new car Vanwall. The cylinder head and valve gear were based on the design of the then dominant Norton 500 cc motorcycle engine, an ex Norton engineer, Leo Kusmicki, having had much to do with it. By 1955 the team had a competitive version of this engine in a well developed car that was beginning to bring some encouragement. Vanwall had reached a point where the engine and transmission had championship potential, but the team was aware that improvements all round would be needed to get a clear advantage. A bold new design was put in hand, with the chassis and suspension entrusted to Colin Chapman and the engine to Harry Weslake.

Chapman's analysis of the car was that to save weight and to improve the handling, a new chassis was needed along with the suspension changes. He produced the outline drawings for both and suggested that Frank Costin be asked to propose a body shape. Frank had by then worked often for Lotus on sportscar bodies. The car that resulted was a quite new shape in motor racing. It was of high build, with a very smoothly-shaped body, wide enough to fill the space between the wheels. It even sported wobbly web wheels. The new car put new heart into the team when Stirling Moss won the Silverstone International Trophy in May 1956 on its first outing, soundly beating the Ferrari team, to the enormous delight of every British racing enthusiast. It was the same Chapman inspired design that went on two years later to win the World Championship – the first time for a British constructor.

Motor sport enthusiasts often refer to the De Dion rear suspension, very widely used towards the end of the front engine era. It found use on many Lotus sports cars and continued until the early Mk 12 single seaters. Other makes used it frequently too, most makers of large sports racing cars included. This view shows the origin. By 1904 De Dion had realised the benefits of mounting the differential separately from the axle. They had also solved the technical difficulties of transferring the driving force from it to the wheels via articulated and telescoping shafts.

The Lotus 30 was a car which caused much excitement at its inception. The largest sportscar Lotus had made had been the 19, with (normally) a 2.5 litre Climax engine. Since that model Lotus had made the Indianapolis car, the most powerful so far, and the Elan, which pioneered the backbone frame. Combining the two ideas into a lightweight sportscar with the power of an Indy racer promised well. However the reality fell way short of the dream and the car was excessively flexible and suffered incurable handling problems. Its successor, the 40 was also disappointing and was unkindly dubbed the 30 with ten more mistakes. Jim Clark is seen at Aintree, 1964.

BRM's story was similar but its problems more extensive. BRM's car had behind it, not the steady progress that the Vanwall had made through different versions, but the legacy of the much admired, and much abused, V 16 car of 1949. This car is remembered for its incredibly powerful engine, approaching 600 hp on 1500 cc, and for its unreliability. Time and again it had promised much but disappointed backers, drivers, mechanics, designers and fans. By 1957 this car had become reasonably competitive but the formula for which it had been designed was no more and the team was by now running a new 2½ litre front-engined design. This car had potential, but was well down on road holding ability as compared with the others in the field. The team was becoming very discouraged with their inability to bring about real improvements.

Tony Rudd, then a key figure at BRM, and later Chapman's right hand man at Lotus, remembers: 'We had competed in a race at Goodwood on Easter Monday 1957 when Roy Salvadori had the servo put the brakes on and could not get the car to the start line. Some of us had stayed at Goodwood and the Tuesday after the race Colin came to drive the one car we had kept there. The Chief Designer Peter Berthon and I were both present – I think Cliff Allison was there too. Colin drove the car for about 10 laps and asked a lot of questions. He agreed the handling was horrible; but was very polite and diffident. He asked for drawings but made no promises and went away'. When Chapman sent his drawings to BRM they showed the use of conventional dampers and coil springs to replace the pneumatic struts they had favoured until then and a lowering of the roll centre. The rear suspension was live de Dion axle located by radius arms and by a block sliding in a groove in the front of the rear gearbox. This meant that the roll centre was close to hub height. Chapman wanted it lowered by the use of a downward extension of the de Dion tube together with a Watts' linkage.

'I was trying to find out which were the important bits – where we had to hit the tolerances and where not. I expressed surprise that he didn't want

to change the front end, because I thought there was quite a bit wrong with it. He agreed it was not right but said it was not as bad as the back – and we would have to make a new car to fix it'.

Rudd also felt that the sliding splines in the rear half shafts were problematic, since under drive torque they would refuse to slide smoothly (Chapman had used the fixed length drive shaft and later the rubber coupling to avoid this very problem). He therefore had BRM make some sliding shafts incorporating ball bearings to ensure proper operation.

'The cars were ready for the French Grand Prix at Rouen and since we had had troubles getting hold of the very top drivers we had done a deal with Lotus driver Mackay-Fraser to join Ron Flockhart for the race.

'On the day Flockhart had an accident on lap three and went off the road on some oil dropped by a Vanwall. The race was Fangio's – he distinguished himself in a Maserati, being so much faster than anyone else. This was in the days of four car teams of Ferraris and Maseratis and drivers such as Fangio, Behra, Musso, Mantovani, Collins and Hawthorn – a tough bunch. But Mackay-Fraser had got up to 5th or 6th place and he was right in with them. For a third of the race his was the leading British car.

'This was not the BRM suspension that we used to know! The whole thing cheered us up immensely'.

ABOVE **One of the failures among Chapman's projects was the type 30. It was nonetheless an exciting concept. Look at the backbone chassis derived from that of the Elan. The main box member houses the fuel bags and the rear legs surround the large American V8 engine. More fuel is carried in the body sides. The engine cooling radiator is in two parts, taking air from each side of the spare wheel in the nose, and exhausting into the front wheel arches.**

ABOVE RIGHT **At an event at Bristol Street Motors' showroom in November 1965, Colin Chapman is seen slot car racing against Jimmy Clark, Jackie Stewart and John (later Sir John) Whitmore. The Scalextric slot cars were first on sale in Britain in the late 1950s. In the early 1960s the author attended Scalex parties organised amongst the Lotus personnel. Someone would volunteer his flat as a venue, and several people would turn up, each with a complete Scalex set. The different sets would be connected up throughout the flat, in and out of the rooms, so that most of it was out of sight of the drivers. This layout (and a crate of beer) made a party. Jim Clark, Peter Arundell and Colin Chapman were among those who attended.**

RIGHT **Tony Vandervell, Tony Brooks, Stirling Moss and Vanwall gathered to celebrate their winning the Formula 1 Manufacturers' Championship in 1958.**

Seat Cover

1958
VANDERVELL
Grandes Epreuves
ANDVOORT · NURBURGRING
O · MONZA · CASABLANCA
the Manufacturers' Cup
ula 1 cars
Vanwall

TOP **This Vanwall might be considered Chapman's first Grand Prix car. Working from an excellent base he transformed the chassis and suspension to produce world beating form. Moss is seen in the car in 1956.**

ABOVE **Stirling Moss again, this time with his preferred number 7. At Chapman's instigation Frank Costin was brought in to design this clean and elegant body form. As with the BRM, the car won its first race following Chapman's re-design.**

Mackay-Fraser was killed the next weekend at Rheims on July 14 and Flockhart was in hospital after his Rouen crash so the team had no drivers.

'So we went to Aintree with Les Leston & Fearless Jack Fairman. I think they broke timing gears and it was a dreadful race – we were really despondent – that was rock bottom in the BRM fortunes.

'While we were at Aintree, though, Jean Behra came up to Raymond Mays and said he had an entry for a non-championship race at Caen the next weekend (28th July) and he would like to drive a BRM. He said that when following it out of corners he couldn't catch it in his Maserati. We were pretty tired – we'd worked all hours to get there because Flockhart's car had been quite badly crashed and now we would have to start and do it all again for the following weekend; it seemed a bit of a shock to say the least. I was softened up by being taken for a slap-up meal at the Adelphi. We just managed to finish the first car on the Thursday night and took it to Eastleigh, put it on one of Silver City's Bristol Freighter air ferries and flew it to Deauville. Raymond Mays had his car and a tow rope and I sat in the racing car and they towed me from Deauville to Caen in the dark – very hairy.

'We met Behra the next day and he practised it – I think he drove it through the streets. We did a lot to it for him to get things right. Anyway we won the race and Harry Schell who had not practised in our second car finished second!'.

These two examples, both of which resulted in first-time wins for the improved cars, bring home to us the respect Chapman could command long before he earned a world renown. They also convey his outstanding grasp of the problems of chassis and suspension design when he had only about seven years of experience in the field. What is more, these were not years spent learning from a more experienced senior; they were years in which he had learnt for himself, feeling out solutions to problems as they were encountered.

Much has been said about Chapman's understanding of the operation of tyres and how to get the maximum performance from them. While building up his knowledge of the engineer's stock-in-trade as a young man, Chapman read two documents which had particular relevance to tyres and their performance. These were a paper by Maurice Olly, an ex-Rolls Royce engineer working for General Motors and a later one by David Hodkin who went from ERA (a pre-war racing car maker) to Ford. The Olly paper was published during the Second World War in the *Autocar* magazine and was the first to produce an analytical approach to the way rubber tyres behave. He showed that given certain data about a tyre, it was possible to calculate its behaviour. David Hodkin's paper was produced for the Institution of Mechanical Engineers around 1954. This one showed how, working among other things from tyre data, behaviour of suspension could be predicted and thus optimised.

These documents provided Chapman with a starting point for much of his work which seemed to others like so much magic.

A third and much better known source of information, which proved invaluable to Chapman, was Ricardo's book on high speed engine design. This was written in the 1920s and became required reading for engine designers for many years. Although only some twenty years more recent than the motor car itself, this work sets out virtually all that was needed for the design of car and racing car engines for more than forty years. It served Chapman well for improvements to Austin Seven engines and right through his life in the specification and development of other units.

4 *The Pros and Cons – Collaborators and Rivals*

IN MOST ACTIVITIES the participants get to know their rivals to some degree. In some – such as wrestling – they come into close contact with them. Motor racing is quite a close community and many of the teams' members know each other despite their rivalry; they share common problems, and despite generally fierce loyalty there are those who move between teams. A sense of community is of course particularly strong among the drivers; last year's arch rival maybe this year's team mate. As for the designers, their rivals are of course other designers and in that circle Colin Chapman was not a person to go unnoticed.

One of Chapman's longest standing rivals is John Cooper, creator (along with his father, Charles) of the Cooper Car Company and designer and driver of many of their products.

Colin Chapman and John Cooper knew each other long before their rivalry on the race track developed from the mid 1950s. Indeed in their early days they sometimes worked together against common problems.

Charles Cooper started to build cars before the war and raced at Brooklands, among other places. After the war he and John worked together with a friend, Eric Brandon, to design a cheap racing car using a motorcycle engine. This, in 1946, was the first of a long line of Cooper racing cars of which more than 1000 were built in the following ten years alone. The Coopers entered straight away into road racing (including speed trials and road hill climbs) and had already achieved some recognition in this sport while Chapman was concentrating on off-road trials cars. Chapman's first road car race was not until 1950 (he was in any case some five years younger than Cooper) and this was in a quite different class of racing.

John Cooper remembers Chapman mainly from the days of the sports car races of the early 1950s, when, as contestants in the same class, they were often pitted against each other. Despite the rivalry, there was clearly an affinity between them. After all, they were both pioneers – in design and in the fact that they were small new constructors battling against longer established companies. It should be remembered that up to their time all successful (Formula 1) racing cars had been the products of the larger concerns with great resources, not to say wealth. Ferrari, Mercedes, Vanwall and others all had large engineering companies behind them with the financial and technical backing this implies. Cooper and Chapman were among the few who were showing themselves capable of giving this type of constructor some competition with little more than a backyard operation.

Coopers' first Grand Prix win came in January 1958 when Stirling Moss won in Argentina. During 1957 they had been running their Formula 2 (1.5 litre) cars in Formula 1 races and caused a scare or two by their performance. By 1958 they were ready to try their hand at F1. Meanwhile Chapman had shown his first single seater, the Mk 12, to the press in October 1956 but had had a rather disappointing season with it in 1957.

In May 1958 Chapman realised an ambition by entering his team for a Grand Prix race – in Monaco. In this race were Cliff Allison and Graham Hill in the two Lotus entries right at the back of the grid, while four out of the front six places were occupied by Coopers.

Naturally John Cooper took an interest in the first Grand Prix offering of his rival but was perfectly confident that the rear engine was the way to go despite feeling that the Lotus was more scientifically designed than his own car and carried many clever ideas. This race was again won by a Cooper and Cliff Allison was the last of the six who finished the race out of the sixteen starters – but it is as much an achievement for a car to finish at all in its first Grand Prix, as it is for the driver.

1958 brought the World Constructor's Championship to Vanwall, but 1959 was Coopers' year; while its significance for Lotus was that it saw the change to rear engine with the design of the 18. The first race of this car in Formula 1 form was in Argentina in February 1960. John Cooper remembers of this race '...that was when Colin Chapman had realised he had got to put the engine in the right place. Innes Ireland drove it very well; it was very quick but luckily for us it was not quite strong enough and so only finished sixth. It was then we realised our car was not going to win any more'. In the event, Cooper redesigned his car 'on the aeroplane going home' and produced the 'low-line' Cooper which won the Championship again, while the fragile Lotus was being worked up to give the reliability necessary.

John Cooper states his opinion of Chapman in no uncertain terms: 'I personally think that Colin Chapman was one of the finest racing car designers

October 1955. The beautifully streamlined Lotus IX seen at Brands Hatch. Colin Chapman is driving (and looking well pleased with himself). Ivor Bueb is seen behind in a central seat Cooper Climax 1100.

of all time – no question about it. His creative ideas were out of this world. Some of his first cars, such as the early single seaters were a bit silly in some ways – they were much more scientific than ours though. But when he produced the first monocoque car it was brilliant. I remember looking at it and thinking that's *It*. It was dead right; everything about it was dead right ... the ultimate.

'The next thing you think of with Colin Chapman was the ground effect car which was fantastic.

'When it came to the twin chassis car he was right again; it was the only way to go if the cars were to get faster and handle properly. The drivers are getting neck and back problems with the rough ride they now get.

'He was undoubtedly the greatest Grand Prix designer – certainly in my time. I got on with him very well although I know many people did not, and my father never really trusted him.'

Over the years that the two knew each other there were many shared experiences. In the early days they often travelled together to races, sometimes sharing a hire car because neither could afford one alone. They flew together quite a bit too, and Cooper, who was a beginner as a pilot while Chapman was already expert, has memories of certain events in which Chapman was either a key player or a participant.

When racing at Nürburgring one Sunday in about 1962, both having race entries for the next day, a Bank holiday, at Brands Hatch, Chapman offered to take Cooper back with him in his Piper Comanche. They were to fly to Biggin Hill with Jim Clark and Innes Ireland to be met by Andrew Ferguson for the run over to Brands Hatch, so it seemed a good idea. On arrival at the military airfield where the plane was parked the weather was so bad that the controller flatly refused to allow their departure. Cooper's reaction was to say 'Thanks very much, I'll go via Cologne as planned'. Chapman said he would soon sort it out and after about half an hour's pressure had talked the controller round to letting them take off. They climbed away through the clouds and only at 7000 feet came into clear air.

Reaching the Channel coast they called for instructions and were told to go down to 3000 feet – back into the weather. This was a strain for Cooper. Although Chapman was a qualified instrument pilot, Cooper was not, but knew enough to be concerned. He said 'do we really have to?' and Chapman replied with what Cooper now thinks was mock concern 'that is what they say'. Above the Channel it was very thick with not even

A younger Colin Chapman is seen in 1956 in the first single seater Lotus, the Mk 12. The simple lines show how the shape was made to enclose the structure of the car in the slimmest possible shape – whereas in its replacement, the Mk 16, the shape was conceived to optimise the aerodynamic performance. Notice also the 'wobbly web' wheels, which were new at the time. They are held on by wheel nuts from the pre-war Austin Seven. The De Dion rear suspension system was replaced by the Chapman strut within the year.

1969. A line-up of cars in the Lotus Components Annexe of the Hethel works: two late Lotus Seven Series IIs – as distinct from the Caterham Sevens they became in 1973, and four Formula Ford single seater Type 61s.

the wingtips in sight, and Cooper became more alarmed when the engine started to run rough. In some anxiety he said to Chapman 'The engine is getting a bit rough isn't it?' and Chapman, laying on his concern, said 'Yes isn't it'. Chapman fiddled with the engine controls until things were better. Cooper has since realised that Chapman knew all along that nothing serious was amiss, but was winding him up over a bit of carburettor icing.

Cooper, of course, had been the creator of the Mini Cooper and remembers a conversation with Chapman at which he said that he was going to do the same sort of thing and make a Lotus Cortina. Chapman asked many questions about Cooper's negotiations over the car and what sort of deal he had made (Cooper received £2 per car).

Mike Costin

Some of Chapman's early collaborators went on to become prominent in their own right. One such was Mike Costin, who left a job at Lotus to go full time at Cosworth Engineering which he had set up with Keith Duckworth. Originally tuning Ford engines for racing, Cosworth became a major constructor of racing engines and is now a multi-disciplined company ranking alongside Rolls Royce within the Vickers group.

In 1952 Costin joined as one of the band of Lotus part-timers who worked so willingly on the early models, the Mk VI and VIII. He retains the memory of when he first met Chapman: 'My first impressions of the bloke were that he was a really switched-on, wide-awake, bright, intelligent and thinking character – that came across with Chapman straight away.

'I started going down after work making parts for the cars and while we were working we often chatted about things to do with the car and the prospects of what Lotus was going to become. Even in those days he had the idea to go on to become a car manufacturer!

'Chapman spent such spare time as he had in research in libraries and so on; looking at what had been done before on linkages and all sorts of things. This helped him particularly since he was a civil engineer moving into this mechanical, automobile field. He did what many do not do enough of – background research. He did not use it to copy though but as a trigger to go forward – that really was his forte.

'His weakness was in design – he did practically no stressing at all. This was where Gilbert Mackintosh came in. He was in the stress analysis department at de Havilland – a stressman, used to doing the intricate calculations necessary for aircraft structures. The chassis of that Mk VIII sports car was stressed by Mac. Every other part of the car that was not stressed by Mac was designed by Colin and was pretty useless; under stiffness and under strength. Colin's use of metal to obtain the necessary strength or stiffness was just not good enough. Chapman had a principle of designing it light, seeing where it failed and then beefing up. This is all very well but it covered up a lack of real understanding of what is required; therefore that is what I would put as one of Colin's limitations. Nobody is perfect and nobody is totally wonderful. You have to remember that Chapman was involved in running a team, screwing the drivers, screwing the companies and people, *and* building cars. Someone like Duckworth on the other hand was only interested in engineering. He wasn't interested in building his company. Duckworth's total interest was on the engineering, where Chapman had it very much spread around and therefore couldn't give the engineering enough time. I am sure that if he had had that time to give it he would have done a good job. In other ways he was brilliant.

'I recall the lunch time when he first showed us the Lotus 25. I think it had been chewing over in his mind but it first came to light in that little restaurant we used to go to in Waltham Abbey. He sketched out the concept that lunch time on a paper serviette and then went away and drew it. Now I don't know whether or not he had actually drawn it before and sketched it for us at the lunch table and then carried on drawing it but the two couldn't have been very far apart. My first impression of what he sketched was that it was perfect because it was aircraft principles without the stupidity of aircraft construction. He built into the structure enough shape and made the skins thick enough so that they did not need interior bulkheads or ribbing. This meant you could easily put the fuel in it. It was smooth inside so you just had to cover the rivet tails and put the fuel bags in.

'The car was the first Lotus to use solid rivets because the pop rivet used before did not fill the holes reliably. The pop rivet clamps the skins together but doesn't really expand to fill in the holes, whereas the solid rivet is forcibly expanded and provides location by bearing rather than by friction. The problems with the undertray of the Seven illustrate the difficulty I think Chapman had. If you take a rectangular structure braced by a pop rivetted panel and deform it, it is only the three or four rivets near the corner that take all the stress initially. It is not until you actually deflect it that the load spreads to the other rivets and by this time the first few have started tearing out – then the load comes onto the next three and so on. But Colin never understood that as a principle. He could not visualise the elastic deformation of a structure. I think this was a weakness and I am not sure he could have understood it even if explained.'

Costin remembers how they worked so intensively – always experimenting – 'learning at a dizzy rate but never having time to consolidate. We were young; we were young and foolish. I was more foolish than later when I was old and foolish'.

Two of Chapman's notable failures were the 17 and the 30 (also the 40, known as the 30 with ten more mistakes). The problem on the 17 was the strut suspension which made the handling quite impossible to predict. Costin narrowly missed the marshal's box at Brands Hatch in a 17 because of the car suddenly going sideways – just because of the road bumps, let alone cornering. The car could enter a corner oversteering and change to violent understeer without warning; another time it would start by understeering desperately and then switch the other way. Despite (or perhaps because of) pressure from members of his team to abandon the strut suspension, Chapman persevered. They tried many tricks to cure the problem – but there was a fundamental fault in the shortness of the strut. The people around Chapman wanted to change to wishbone suspension because at least that was known to work and be manageable but Chapman stuck to his guns and the car went into production with the strut suspension – although later they all had to be changed because they were undriveable. On the Lotus 30 the fundamental problem was that the chassis just was not stiff enough for that weight and power – it was an extension of the Elan backbone chassis principle but was not up to the job with this type of car.

Chapman's adoption and development of ground effect was really a first. The ground effect car works, put simply, by the front of the car scraping the air off the road as it goes along, leaving a partial vacuum underneath it. Of course, to make it work well you need side skirts and you can also get further benefits by doing clever things with the shaping of the underside. The ground effect idea was the saviour of Cosworth.

Ferrari was running a flat 12 cylinder engine which was comfortably beating the V8 Cosworth powered Lotus which had been so successful the year before. One of the key reasons was that a flat engine shape allowed for a lower rear body line and left a clearer passage under the wing as compared with the V engine form. This meant that the wing was more effective. Cosworth feared that it would have to change its engine design to emulate Ferrari and give Lotus the same advantage, but Cosworth could not have afforded the cost of a totally new engine project at that stage. Cosworth was therefore putting off the evil day at the time when Chapman produced his magic. Magic indeed for Cosworth because a V engine is just what is needed for a ground effect car since it accommodates the duct-like form of the underside. The new Lotus turned the tables completely and it was Ferrari who had to make a new engine.

Despite their long collaboration, Costin says that he and Chapman were never friends – it just was

Introduced in the hope of building on the reputation and sales of the earlier car, the Series IV Seven (1973) did not quite capture the atmosphere that would have allowed it to have been a comparable commercial success. The body is in glass fibre over a steel chassis.

not that sort of relationship. Their partnership lasted over a period of ten years working together daily and a further 20 years in which they were close business associates. However, no more than a working relationship ever developed.

Gordon Murray and Mauro Forghieri

At the time of writing the most successful racing car designer is Gordon Murray of McLaren and formerly of Brabham whose designs have dominated Formula 1 for a decade winning multiple consecutive Constructors Championships.

His involvement with racing car design goes back some 20 years to 1970 when he came to England wanting to get into car design, but not daring to hope that he could get into that of racing cars, let alone Formula 1. As a teenager in South Africa, Murray had idolised Chapman and the Lotus cars he designed. He had built his own Clubman's class racer fairly close to the Lotus 7, but incorporated all his own ideas. Like so many aspiring special builders he studied the Costin and Phipps book *Racing and Sports Car Chassis Design* (of 1961) and devoured *Autosport* avidly every week.

Gordon Murray still finds that the Lotus Seven and the Elan are two of the most perfect cars ever conceived. In fact during his first year in England he blew practically all his funds on an Elan. He had hoped for a job with Lotus, but they were going through a bad patch and were shedding jobs rather than hiring people. By the end of six months of looking for a way in to the racing world, he so desperately wanted to possess an Elan that he bought one before the money he had saved ran out entirely – still with no job.

Murray is now of course very much established and is able to run the car for fun rather than for daily transport, in fact he has bought another Elan – over twenty years old – just because he likes them. His liking for Lotus cars is not, however, a blind love. He frankly says that they were not much as production cars – too many design and development faults – but finds them conceptually brilliant.

Murray's view of Chapman is very much coloured by the feeling that, particularly in the later years, he spread himself too thinly, by trying to be active in so many fields at once and therefore perhaps did not fully live up to his promise.

Chapman's view of the younger Murray can be

BELOW **The Elan Plus Two was introduced to lead the potential customer up-market. The person who bought an Elan some years before, now married and perhaps with small children, might be enticed into replacing it with the new model with the extra space.**

ABOVE RIGHT **The sign-off: a latter-day Elite on the rolling road, undergoing final checks before release for sale.**

RIGHT **Some Elan Plus Twos and two Elans nearing completion at Hethel, 1968.**

CPW 528K

ABOVE **Europas on the assembly line at Hethel.**

ABOVE LEFT **Right hand side of a press Europa. The side window frame looks a bit mis-shapen.**

LEFT **A three quarter front view of the same car. The front bumper is from the contemporary Ford Anglia.**

assessed in some measure by something that happened after the South African Grand Prix of 1973. At the time Murray was designing for Brabham and was in fact in virtual sole charge of the Formula 1 team. Up to this point Murray had never met Chapman, but had seen him many times at race meetings – although he says he would never have dared to go up to his idol and speak to him. In practice the Brabham team cars had been first and second on the grid, they had finished first and second in the race and made fastest lap; and you cannot do better than that. In addition another lent to a private entrant and driven by John Watson had finished well up and the three cars had at one stage been running 1-2-3. Finishing the season in this spectacular style caused Chapman to come up to Murray and say 'That is the way it should be done!'. Following this event, Murray received a job offer with Lotus, through a third party, at a salary which really made him think about it. However he preferred to stay where he had control, rather than go where he felt he would have to play second fiddle – even to an idol. He is nevertheless understandably proud of the fact that he received similar job offers almost every year after that first one.

Murray has never ceased to admire Chapman and, as for so many others in and around racing, the Lotus 25 is for Murray the essence of Chapman's genius. Its simplicity – indeed the way it brought simplicity – was extremely elegant, and as a solution to the problems facing the grand prix car designer at the time it was well nigh perfect.

Murray is remembered also for the Brabham

'fan' car of 1978. This was a desperate move on his part to compete with the then dominant Lotus 79 ground effect car which Cosworth had been so pleased with. Brabham, like Ferrari, had a flat engine; theirs was an Alfa Romeo unit. For some time the rival teams had agonised over Lotus' magic ingredient, because of course the operating principles were not published. Murray had guessed what went on under the Lotus and says it came as a nasty change to be in a position of having to copy someone else; for some time now it had been his designs that caused others to copy. To work out why the Lotus was so fast was by no means to have the way to match it. Murray realised that he could not make a true ground effect answer because of the engine shape. He toyed with ideas – even thinking of separating the engine and gearbox so as to fit the necessary under-car ducting in the space so created; he pored over the yellow F1 rule book until it was dog-eared, trying to find something to exploit to get back in the running.

Finally, with a stroke of his own genius he united the notion of an engine-driven fan as used at one stage on the Chaparral with the words in the rules 'primary function'. The rules disallowed moving components whose primary function was as an aerodynamic aid. But a fan whose primary function was to cool the engine was obviously another thing. It would be quite normal to draw air for cooling from under the car of course. So the fan car concept came together: the engine driving a fan whose *primary function* was cooling but which also drew the car down onto the road.

The whole idea turned around the definition of those words primary function. But there was no definition given in the rules how the primary function of something should be determined. This was the loophole needed. Study of dictionaries showed that in the absence of anything else it could be said that if a component served more than 50% for some function this must be its primary function. Now a fan of which say 40% of the effort went into sucking the car down could be really something. And so it proved. The car design was rushed through in three months and had its first outing in Sweden. Murray recalls 'It was so fast through the corners that to hide its performance from the others as much as possible in practice we were running on full tanks and so on to slow the car's lap speeds'. In the outcome the car won the race – basically if it kept going it would inevitably win; but it is none-the-less unusual for *any* new car to win its first race whatever advantage it might have.

Obviously such a move did not go unchallenged and Chapman was not the last to speak up. He recognised the potential more clearly than most and approached Murray to persuade him to drop the idea. He reasoned – why do you want to do this; we will all have to do it: is it the way we want all Formula 1 cars to be for ever? He tried bullying and cajoling Murray with talk of the sporting spirit of Formula 1. Also, Murray believes that Chapman encouraged a group of drivers to say they would not race against it because it threw out stones. The authorities measured the car and found, sure enough that more than 50% of the air from the fan was used for cooling; thus the 'primary function' rule was complied with. The car was declared legal but this did not stop pressure being applied by the members of FOCA, the Formula One Constructors' Association, such that the car did not race again.

It is interesting to consider how Chapman felt when he was on the side of the protestors against an innovation. All the arguments against the fan car might have been used against him at other times. One can also wonder whether Chapman failed to realise that although perhaps less that 50% of the fan's power went into helping the car, the overall size of the fan may well have been several times the minimum necessary for cooling the engine – indeed with Formula 1 cars there is no shortage of air flow for cooling. The designer considering using fan cooling, without calling upon it to improve cornering performance as with the Brabham, would be weighing the power loss represented by the air drag of a conventional system against the power required to drive a fan cooling system. In order to gain anything from the idea it would be necessary that the fan consumed less power than did the unassisted cooling more normally used; therefore the fan would be as small as possible.

Murray remembers the case of the Lotus twin chassis Formula 1 car. He feels the balance of justice is slightly different because of the greater complexity of the Lotus concept. He found that his admiration for the idea as a solution to the tech-

Stirling Moss in a Lotus 18. This car was entered by Rob Walker for the 1961 Belgian Grand Prix at Spa where it finished eighth. In a similar car the year before Moss had a severe accident during practice. The accident was caused by the fatigue failure of the rear axle shaft flange, resulting in the wheel coming off. This was a design defect that certainly should not have happened. Moss made a miraculous recovery from two broken legs and crushed vertebrae, and was racing again in less than two months.

nical problem, and his admiration for Chapman, meant that some part of him wanted the car to be accepted as legal. On the other hand he was against it because of its effect on Formula 1, and Brabham in particular, in forcing all to follow.

This re-introduces an important aspect of the question of 'the establishment', apparently so often closing ranks against Chapman – and as we have seen – against others. Chapman was very strong in his argument in favour of technical freedom (within the formula) for the designers. However the counter-argument is that the formula exists to maintain the shape and character of the sport, just as do the laws of tennis or football. If every technical innovation that could worm its way through the loopholes of the rule book was allowed, then by now no Formula 1 car would be worth entering in a race if it did not have, say four front wheels and four rear wheels, all driven; an aerodynamically effective outer chassis aided by a separate gas turbine to hold the car down, and maybe a system for spinning up a flywheel during braking so it could aid subsequent acceleration; the whole being controlled by a super-computer at the constructor's headquarters via a satellite link. It is thus to keep the whole thing within bounds that the rules are made and from time to time undergo redefinition to constrain runaway technology at the hands of genius within design offices. The real difficulty before those involved is therefore not whether to adjust the rules from time to time but the individual decision in each case as to whether the innovation lies outside what is good for the sport or not. There will always be controversy at these moments.

Murray has very few adverse points to make about Chapman. They are mainly to do with the detailed execution of his designs – often done by others, but for which he was nonetheless responsible. Murray feels (as did Mike Costin) that if he had had less diverse interests – or perhaps had people looking after them to whom he could hand over more completely – then he would have had even greater success in Formula 1. Chapman was one of the very few whole car designers to make his own place in history. He may be one of the last, since the era of the one-man design team is over, due to the diversity of the technologies used.

Mauro Forghieri, now leading Lamborghini's racing efforts, worked at Ferrari for a total of twenty four years up to 1984, for most of which time he was responsible for the design of their formidable racing cars, carrying on the tradition that made Ferrari the evocative name that it is. He first met Chapman on a flight to Nürburgring in 1964 (Lotus having won the World Championship the year before).

Forghieri's view of Chapman is of course more remote than Murray's – from another country for one thing – but his impression was that Chapman was perhaps more dedicated to performance than he was to reliability. He recognises him as a great innovator and while not having consciously copied his work, says that one is influenced by what one sees as a designer. Inevitably an engineer is always turning over his design problems in his mind. This means that the mind is very aware of current difficulties and is consequently able to absorb relevant design solutions executed by others when encountered. Such solutions will frequently show the way forward in a new design, without this constituting copying. Having said this, there are plenty who find copying the ideas of others quite an acceptable way to work.

Forghieri sees the twin chassis concept as a sound approach, ahead of its time, but would rate the monocoque chassis together with its suspension system as Chapman's greatest contribution to the Formula 1 scene.

Chapman's foray into turbine driven cars was one of his failures, being too innovative – too much all at once. Here was another case where Chapman's perseverance perhaps worked against him, since he would possibly have done better to have abandoned the idea sooner.

The number of times Forghieri met Chapman was fairly limited, but he says he unfailingly received from him great kindness and consideration. He rates Chapman as one of the best ever car designers, adding that he had always heard well of his caring attitude towards his team.

Stirling Moss

Stirling Moss, among the finest of racing drivers in living memory, was a rival to Chapman in that he often raced against Team Lotus, but in a Lotus. Perhaps ironically, Lotus' first Grand Prix win was achieved by him, while the works cars retired. Moss was, in many ways, uniquely touched by Chapman's influence.

He had always wanted to drive British cars, but until the Vanwall reached a good state of development, no British car could hold its own in international competition. For all the years following the war, British enthusiasts had hoped in vain for a Grand Prix car capable of offering a serious challenge to the established makers in the field. The first British car to win a Grand Prix since the crea-

tion of the World Championship was the Connaught when Tony Brooks brought one home first in the non-Championship Syracuse Grand Prix of 1955. Winning this event was like the breaking of the four minute mile barrier; it showed it could be done, and reinforced the resolve of the contenders and the patience of the fans. From that time on British Grand Prix constructors undertook the long climb to dominance. Towards the end of the 2½ litre formula there were many British makes capable of winning, including Lotus with the 18. Moss's first drive in a Lotus was in the Monaco Grand Prix of 1960 where he won the race in Rob Walker's car.

Why Stirling Moss might be thought to have been particularly affected by Chapman was that he suffered failures in the cars which caused him accidents. The first accident which was surely attributable to a failure broke his legs and back.

One might expect that Moss would not be too ready to speak favourably of Chapman, but in fact he is very quick with praise, while maintaining a

For 1961 Lotus produced the 21 for the team. Stirling Moss, driving for private entrant Rob Walker, could not get hold of this latest model. Walker therefore had the chassis of his 18s modified and new bodies made to produce the so-called 18/21. Here Moss is seen on his way to winning the German GP at Nürburgring, 6th August 1961.

cool understanding of the fact that Chapman was indeed responsible. In his own words: 'From my point of view I could see that Colin was a genius although he hadn't got much of a sense of humour. I think he had extremely good innovative ideas; each time they were unique to him and they definitely worked. Having said that I think his main problem was that the design he did was aimed at a very narrow band of people. If you were really good as a driver a Lotus design was brilliant. If you were not very good, it was impossible to exploit the benefits you could get from the car.

'If you were a moderate driver or even a good

In this view of the same car can be seen the car's Mk 18 suspension details, giving away its origins.

driver with a Cooper you could get away with it; you could do anything you liked because it was a blacksmith thing and it worked wonderfully well and did everything it should do and it never bit you. I think that probably the lateral 'g' you could generate on a Lotus was better than on a Cooper but there were very few people who could get up into that echelon; getting there was difficult and the experience was not fulfilling except in achievement. I never had the pleasure of driving a Lotus other than finishing the race in the lead. That is to me where the pleasure started and finished. It was not a fun car. Many others, Maserati, BRM, have their own pleasure.

'Now the pleasure and exhilaration of driving a Lotus was only there for a fleeting moment of success. It was a tool with which to win in the right hands but it was a car that I was always worried about, it was waiting to bite you. You could liken it to going out on a date with a bad tempered girl. If you made the wrong move at the wrong time you may get your teeth knocked down your throat.

'That does not mean I criticise it, it means that really he was brilliant in what he could achieve but the way that he did it was not very usable by most people. The cars, I think, were built too light – I don't know how many wheels I had come off but quite a lot. They came off because the stub axles broke; not because I hit anything but because, thanks to Chapman's ability in generating the lateral 'g', of the loading I put through the stub axles.

'One is only talking about very small amounts, because the differences between that and the Cooper were not that high but you are talking about a small percentage.'

Arising from the numerous failures Moss suffered on the Lotus cars he drove so well, he was on one public occasion moved to put Chapman's temper to the test. Following his win of the US Grand Prix of 1960, Moss was invited to a party to celebrate that and his birthday, which fell at the same time: 'I reckon Colin didn't have a very good sense of humour, because I won the US Grand Prix that year and they had bought me a cake because it was my birthday. The cake had a car, my Lotus, on the top so I cut the front wheel off and said I would like to give the first cut to Colin Chapman. He didn't think it was funny at all'.

Moss is generous in his admiration of Chapman, despite the problems that he suffered: 'There are no two ways about it he was a brilliant designer but with limitations.

'His strength was in innovation, new ideas, his skilful way of getting what he wanted mechanically; but his weaknesses were first that frankly he didn't seem to have the ability to make it easy to utilise his skills – his cars were too difficult to drive; and secondly that from what I could see he was prepared to pare the things down to the bone without leaving either sufficient safety margin or doing enough research to cut out problems. I think he took short cuts. He was perhaps a bit lax on fine detail.

'About the accidents, I feel no animosity but I certainly think he should not have done it, a man of his knowledge. He had no right for it not to be right on the dangerous things. At one time, Rob Walker, whose team I drove for, was changing my stub axles and I think the rear drive shafts for every race and even then they broke. That shouldn't have been necessary. After all that I do have more respect for him as a designer than I have animosity for what he caused.

'I quite liked the man but it is very difficult to look down a corkscrew straight – he had devious ways. I think if I was going to do something with Colin I would write and confirm it. There are people like Bernie Ecclestone and John Webb who were very tough, difficult people. If they said something on the phone I would take it as fixed, but with Colin I don't think I would have done. Maybe I am being unfair in saying that, but my general feeling was that maybe he means it but let's confirm it. I think he could be quite devious – he could think round ahead of you.

'Socially I got along with him alright; he wasn't a close personal friend but having said that I quite liked the guy; I had an admiration for his skill and he was a good driver too. I wouldn't say we were particularly close because Colin wasn't the sort of person that one would normally get close to. I would say that Colin and I had quite a reasonable relationship. If we were at a party together we would talk but he wouldn't have me to dinner and I wouldn't particularly have him to dinner.

The great thing about Colin I suppose is that he did know quite a lot and he had quite a lot up top. The way we used to flounder around at Maserati, particularly on the aerodynamics side was unbelievable. Colin was not an aerodynamicist but he made sure he had access to them. He had good people working with him as well, Costin and so on. These people were not stupid and I think that he benefitted very much from that fact as well. I think in his way he was a genius.'

Collaborators

Aerodynamicist Peter Wright, now a director of Team Lotus, was the chief expert behind the ground effect research and the twin chassis concept. He was recruited into Lotus from BRM by Tony Rudd in 1974 and worked to start with on the boats. He had, however, been dabbling with the ideas behind ground effect since 1969. Some time in 1977, Chapman moved him to Team Lotus and put him to concentrate full time on making the ground effect system work.

One of the ways in which Wright found Chapman good to work for was that he knew when to back an idea and when not to. Although he rejected many, when he did back an idea he did so wholeheartedly and did everything to support those involved, better than almost anybody else in industry. If he thought an idea was valid and worth doing, however hard it was, he would throw everything behind it with a total commitment. Accepting that many new ideas are very hard to make work well, Chapman knew that many people resist innovation, but he loved it.

Wright's views of the Brabham fan car are, not surprisingly, different from those of Gordon Murray. When analyzing that question of prime function, Murray said, correctly, that more than half the airflow was aimed at engine cooling. Wright, on the other hand, points out that a fan in engineering terms is a device for producing a pressure difference. By applying the pressure difference to some object a force is produced and that force is proportional to the area of the object concerned. In the fan car the force acted in two places; on the radiator where it created air movement, and on the underside of the car where it produced a down-force. The area of the radiator was about one twentieth of that of the underside of the car. Therefore only 5% of the fan's function was going into the cooling, not 50%. The whole question is much more complex than can be gone into here in proper detail.

Of the various people talked to who worked in the same field as Chapman during his life, none offered a really bad criticism and most expressed considerable admiration. Perhaps one of the more telling reactions, though, came from his own son, Clive Chapman. On Chapman senior's death Team Lotus became the joint property of the surviving members of the family. Clive has the role of Commercial Manager and is still filled with wonder at having been blessed with such a legacy. At the same time, however, he misses the ability to turn to his father for the guidance and support he always experienced during his lifetime.

5 *Suspension Evolution*

THE EVOLUTION OF suspensions was looked at from some points of view in Chapter 3, but it is now perhaps worth a look not so much at the effect of suspension on road holding, but at the different ways they can be physically realised.

Suspension members take different forms with different designers. They can be designed so that the material from which they are made is loaded in any of the possible ways – tension, compression, bending, shear or torsion, or indeed any combination of these.

As has been shown, the most efficient and therefore the lightest structures are produced when the material is loaded in tension, compression or shear. Chapman, always the purist, naturally aimed for efficiency in anything he did, only permitting other forms of stress when it gave some particular advantage, or when to avoid it would itself have created disadvantages.

Since the wheels of a car are carried at some distance laterally from the suspension mountings and generate fore and aft loads, they actually impose bending loads on the suspension system that supports them.

Bending loads can be converted into a combination of the preferred stresses by thoughtful component design, as in the chassis frame. This is done primarily by having a tension member and a compression member working as a pair. The actual forces that act upon them will depend upon the distance between them – the disposition of the material. The wide spread wishbone systems used by Chapman carried this to a logical conclusion, reducing the forces to a minimum.

Generally we think of a typical Lotus system as having a wishbone and a radius arm, although geometrically they are identical to a wishbone of which the base is that of the outside dimensions of the whole assembly. This style of suspension was not solely the province of Chapman but he was one of the most successful exploiters of the type.

An interesting sidelight on the use of the drive shaft in the rear suspension arose in the 1950s, when the fixed length shaft was used as a suspension member. The makers of the shafts, finding themselves asked to supply shafts on which the safety of the car depended much more directly than they were used to, insisted on installing safety rods through their centres. These rods were 5/16 inches (8 mm) in diameter and retained at their ends by lightweight circlips. Few of the car constructors believed the rods would do any good if the tube itself failed since they were only one sixth of the strength of the tube, even assuming that the clips held.

Looking at the suspensions of Lotus cars up to about 1961 it can be seen that they have a fairly restricted steering lock. This was done on the basis that in racing the steering is never turned more than a little. At racing speeds on the normal circuit and with a good-handling car this is true – a steering wheel movement of half to three quarters of a turn is about all that is needed – as long, that is, as everything is going well. Chapman's view at this time was that if the driver needed more lock angle than this he was out of control and it was too late anyway.

Chapman took advantage of this principle in the design of the sports-racing cars of the period to have close fitting bodywork, thus reducing the frontal area. Also the Mk 16, although a single seater, had its body reaching out close to the front wheels to reduce front-end drag.

Later Lotus models have much greater steering angles available, such as to do credit to a London taxi.

But what brought about this change of view? There was at the time a Dutch racing driver by the name of Rob Slotemacher, who, although not right at the top in racing, made a name for himself with his incredible skid control. His party piece was to cross the finishing line in a high-speed spin, keeping the car travelling along the centre of the road, stopping the spin with the car in line again and continuing on his way.

Slotemacher made such a study of the control of a car at extreme attitudes that he eventually started a skid school. This was not a school for the average driver to learn to cope with ice, but attracted some of the top people in the racing world to extend their skills and understanding.

Chapman went along and came back with a new principle – it is never too late. From then on all Lotus racing models had at least forty five degrees of steering lock.

As if in vindication, in team manager Andrew Ferguson's office there used to be a photo of Jim Clark in a 25 with the car almost at right angles to the line of travel, looking over his shoulder while applying its very full steering lock to straighten things out.

It has been seen how one of the requirements in making a successful strut suspension is sufficient distance between the sliding bearings in the strut to ensure that they work correctly. This in turn means

The car on the left has a simple beam axle. Here the loads acting on the wheel mostly result in bending stresses in the axle itself. Thus the structure of the beam will be the heavier since the material is being loaded inefficiently.
The car on the right has a wishbone style of suspension in which the loads are reacting only as tension and compression in the members. Generally speaking the more widely they are spaced at their chassis attachments, the lower will be the loads in them and consequently in the chassis pick-up points. This form allows for the lightest construction.

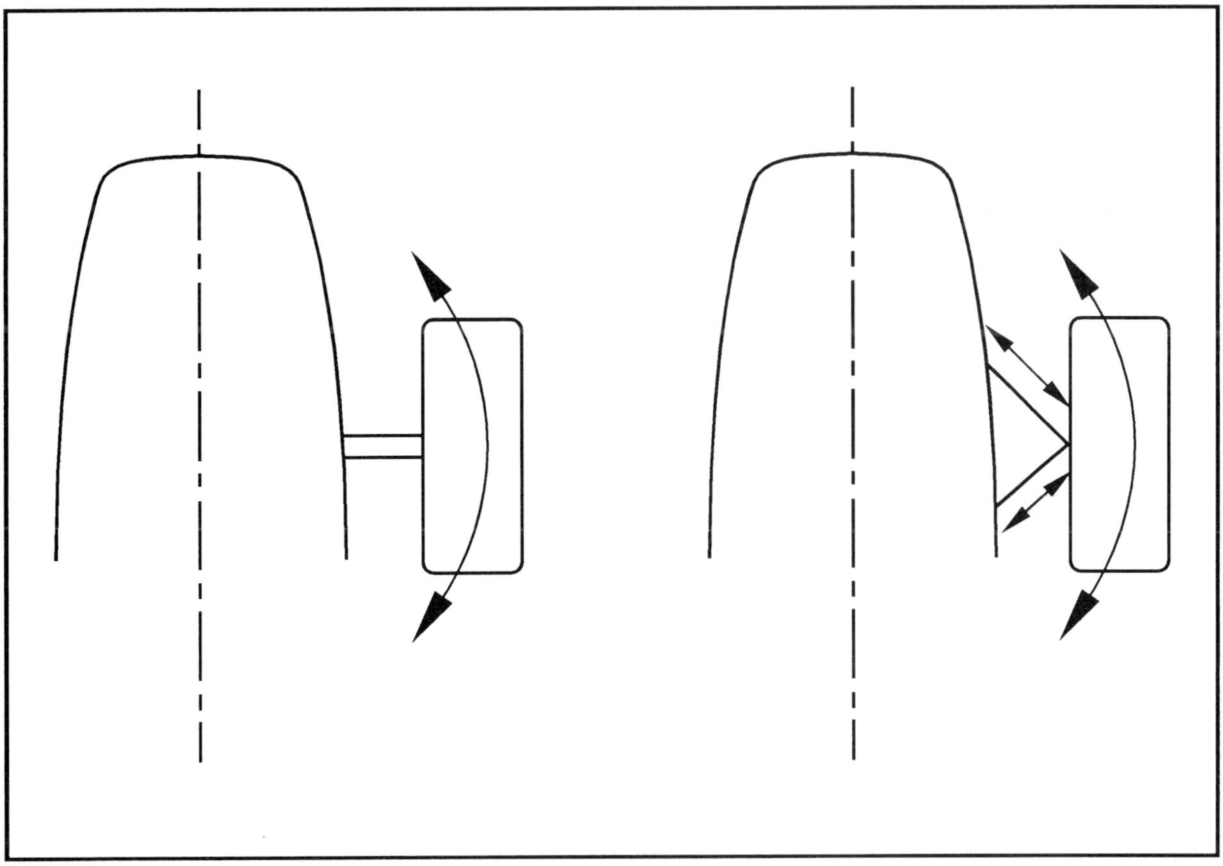

that sufficient height is needed in the car structure to accommodate this length. For this reason, struts are at their best in road-going cars where this height is most readily provided.

Chapman's most successful strut designs are to be found in the Lotus road sports cars. Of these, the first was the Elite of 1957 in which the strut was combined with a fixed length drive shaft. This was followed by the Elan in which it was used instead with rubber driveshaft joints. The reasoning which led Chapman to use this type of drive had as its starting point the problem of 'stiction' in the then more normal splined driveshafts. The fixed length shaft overcame this problem but brought with it certain constraints in suspension geometry, since the only possible attachment points are the output point from the final drive unit and the wheel shaft on the suspension. For a long time, Chapman had wanted to meet the particular requirement for a variable length shaft without sliding friction.

One of the first solutions tried was on the first Indianapolis cars, where a splined shaft was used with ball bearings in the splines, thus almost getting rid of friction. This design was successful but its drawback was its expense, making it therefore only acceptable for a racing car.

When the Hillman Imp was produced it made use of a rubber joint of a type previously used only in industrial drives. The fact that it had been adopted by a volume car producer meant that its price and availability both improved. Chapman saw that this joint could offer what was needed, since it could accept an axial movement as well as the angular misalignments.

The first installation at Lotus of this type of joint was on a prototype for the Elan. At this stage in its life the car was to have inboard rear disk brakes for minimum unsprung weight. With this system, which was of course well established, the braking torque is transmitted from the wheel to the brake

A view of an 18 outside the Cheshunt works serves also to show the field behind where later the Lotus Cortina shops were built. The same field had previously been used to store the Elite body shells awaiting build.

via the driveshafts. The introduction of the rubber into the path of the brake torque had two side effects, one understandable but not foreseen, the other curious and perhaps a chance result of a combination of factors.

Of these, the first was that when the car was parked it was not anchored firmly. This meant that with the handbrake on the rear disks fully applied, the car could be moved an inch or two forward and back. Although this was not a problem much of the time, it could be embarrassing parking down a slope, where the car, having been stopped on the footbrake, would move forward when the handbrake was applied and the pressure removed from the footbrake.

The other side effect could even be regarded as an advantage, since it was an early warning of an icy road surface. It was found that if the road surface had a light coating of black ice, a very light application of the brakes – barely more than a caress of the pedal – would produce a rattling in the differential as an oscillation occurred in the output shafts. This effect would happen without any suggestion of wheel locking and could therefore be heard by the driver before he found out about the ice in another way.

Subsequent versions of the Elan abandoned the inboard brake in favour of one mounted on the suspension upright, which, for packaging reasons, was a sort of 'inboard/outboard' installation.

A little known version of the Elan rear suspension was that fitted to the Lotus Cortina. Production cars had a live rear axle, but two were built with independent suspensions.

These two cars used Elan differentials and struts. The first to be built was purely experimental and had the upper mounts of the differential mounted in a sheet metal 'bathtub' let into the thin steel floor under the rear seat. As this area of the floor was one of the points at which the wheel torque was reacted, the wave-like deflections of the panel had to be seen to be believed. The second of these cars was built for Jim Clark and had the differential mounted on a stout cross member attached to the subframe, independent of the floor panel.

Below the differential was a frame rather similar to the lower rear extension of the Elan chassis. This reacted the differential torque and provided mountings for the wishbones, which were enlarged and strengthened Elan ones.

The subframe was mounted at its front end to the transmission tunnel and at the rear by two arms reaching out and up to the underbody reinforcements, the whole being removable. Modified Elan handbrake parts were used and special rear hubs were made to carry the Lotus Cortina wheels.

At this time these experiments were highly secret but the cat was let half out of the bag one day when there was a factory tour laid on for a group of journalists. They were not shown the test car, but Chapman took one or two of his friends among them to lunch with him in the car he was using at the time – the first of these two Cortinas. The passengers in the car may not have felt the difference, but those waiting for other transport could not fail to see the way the rear wheels tucked into the wheelarches.

Inboard front suspension

The design of front suspension took another step forward on the 21 when inboard spring units were used for the first time. The reasons for this development were the reduction of effective unsprung weight and the reduction of the aerodynamic drag produced by the front suspension units, previously in the airstream. It should be remembered, however, that they were only in the airstream because the now increased steering lock demanded it.

An inboard spring version of the double wishbone suspension needed a form of lever protruding inward from one of the wishbones to which to attach the spring units. Chapman opted to use the upper wishbone to convert to a lever because this posed less consequential difficulties.

Remember that materials are not very efficient in bending, which means that a relatively complex structure had to replace the conventional tubular frame previously suitable. Chapman's scheme for this started with a fabricated arm in two halves. Each half was a tapering flattened tube made up by hand from sheet steel and was welded to a heavy steel tube which formed the pivot, being carried on needle roller bearings. The outer half of the arm ended in a housing for the upper ball joint of the steering upright and the inner half ended in a lug for attachment of the spring/damper unit.

Bearing in mind that all this was at the time entirely new, no accurate figures were available for the loads involved, nor for the strength that would result from construction. The loads were of course estimated, but it was not possible with the information available to make anything more than a rough calculation.

A prototype arm was made and taken to Enfield Technical College's materials laboratory, where testing machines were used to determine the be-

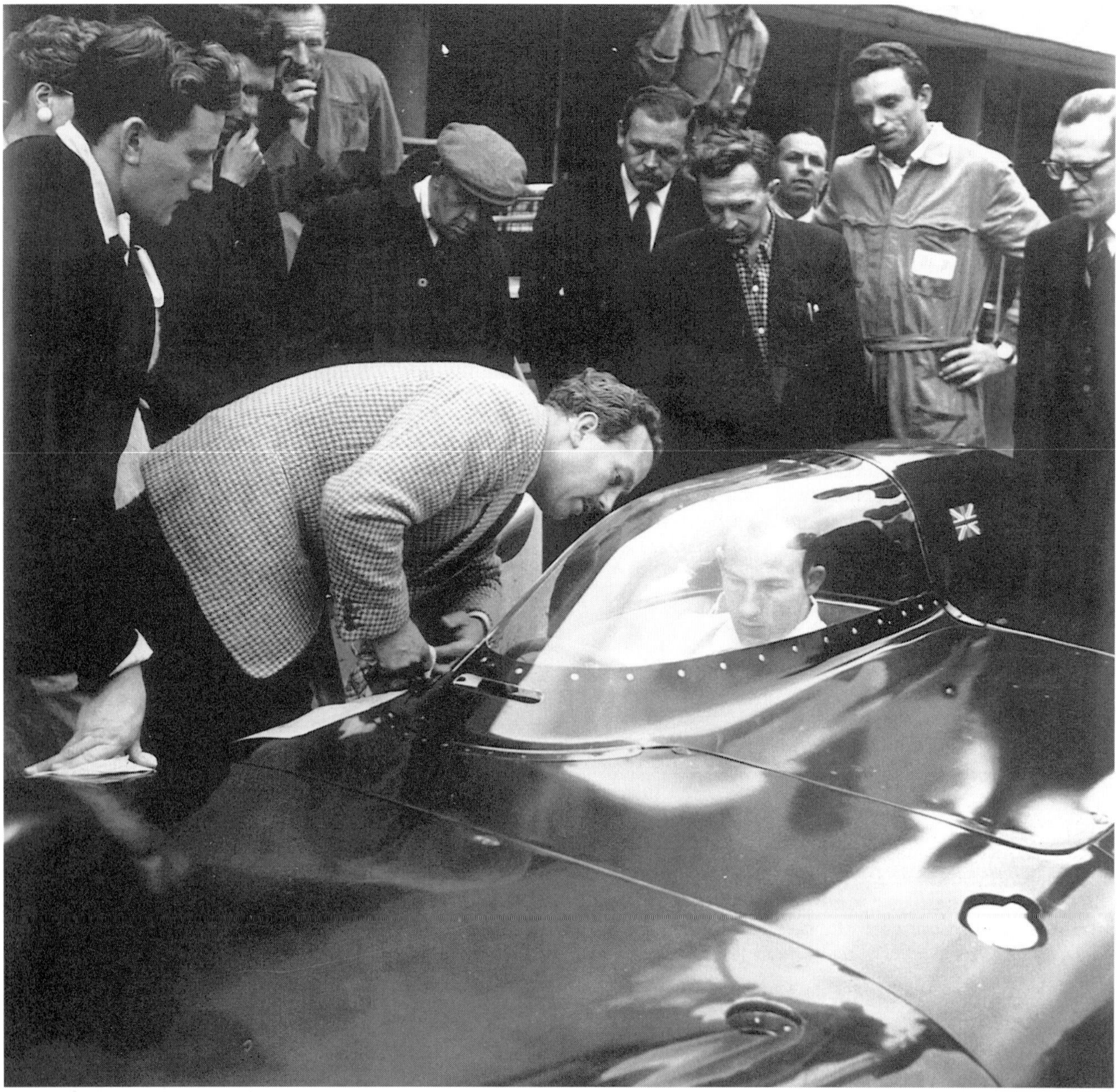

haviour of the arm when loaded as on the car up to the calculated maximum loads. It was found that the outer part of the arm buckled before maximum load, so another sample was made with an internal bracing web. This version withstood the fairly simple test and gave the team enough confidence to build a pair to fit to the test car.

Up to and including the development of this new suspension, the steering rack had been mounted at a height between the levels of the two wishbones.

Stirling Moss is seen being shut inside a special Lotus Eleven in which he set a number of records at up to 135.5 mph (218 kph) at Monza in September of 1956.

This meant that the track rods too were in the airstream. Removing the spring unit from this position made it desirable to remove the track rods as well. This was done on the 24 by raising them until they lay in the plane of the new top arm.

The track rod was made the same length as the wishbone and the whole system fitted with precision bearings, the 'Rose' type joints of the linkages complementing the needle roller system used for the top arm. This resulted in a very precise steering layout with no bump steer effect.

A further result was that there was no Ackerman effect either. This was entirely in line with the thinking of the time which was that on a corner there was no point in turning the inner wheel to a larger angle than the outer, since the inner wheel has little or no weight on it due to the weight transfer on to the outer wheel. The inner wheel, as it cannot contribute to cornering force, is effectively a dead weight and drag and should therefore be aligned as well as possible with the airflow – thus zero or sometimes negative Ackerman effect. (Remember that at this time wheels were narrower in proportion to their diameter than we are used to seeing today, so that alignment with the airflow was more significant.) Once ground effect cars came in the weight transfer was made less significant because the aerodynamic download ensured that there was plenty of weight on both wheels. So the balance of aims and constraints changed again.

The upper suspension arm became more of a precision component than before, and also, because it was now carrying the car's weight, was subject to much greater loadings. Consequently it was made more rugged by taking out the adjusting threads that had always been a feature of Chapman's top wishbones. Adjustment for camber was therefore incorporated in the lower wishbone, which was still of fairly conventional design.

With this revolution in the front suspension links came the first abandonment of the forged steel front upright. This component had served well on all Lotus designs since the move away from swing axles in the 1950s, apart from those using struts. The upright came from the Standard Eight and Ten (the latter sold in the US as the Triumph Sedan), among the first British small car models to

This view of an 18 shows the extreme lock angle provided after Chapman's visit to Slotemacher's skid school. Note also the saddle fuel tank, the water pipe from engine to radiator (insulated where it passes the driver). The gearchange is on the left operating the positive stop shift of the Lotus transaxle unit. Through the wheel are the tachometer, oil pressure and water temperature gauges. By the driver's left hand is the fuel cock.

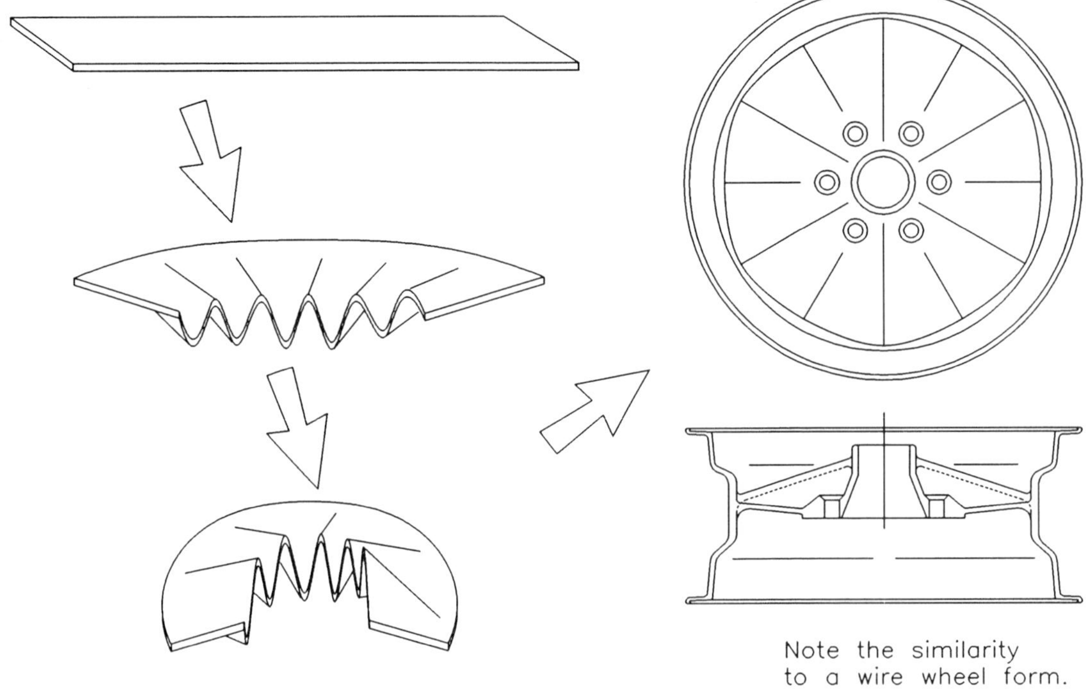

How to visualise the wobbly web wheel.

The wobbly web wheel was really a very elegant concept. At a single stroke it meets all of the basic needs: the centre disc is very strong radially (load carrying) and laterally (cornering). It has its maximum bending strength at the hub, where the bending load is greatest and at the same time the web spreads the loads into the hub over a wide base. One disadvantage is the absence of air flow through the wheel, although with inboard brake location this is not a problem (See p.67).

have independent front suspension. It was widely used by racing car manufacturers including Cooper and Lola as well as Lotus.

Originally used more or less as produced, this upright was progressively adapted as the need for precision in the system grew. A modified version of the steering arm for this upright was used in conjunction with it, as was the wheel spindle (stub axle) which was taper-fitted to it and held with a nut.

At the time when Lotus (and also Cooper) were scaring the 2.5 litre Formula One regulars with their 2.2 litre Climax engined cars, but were not yet taken as serious contenders, BRM were using a front suspension upright designed for the purpose. This upright was in heat-treated steel and was machined all over to a fine finish. It was in the form of a complex channel with the stub axle formed in one piece with it. The interior of the stub axle was also machined out for lightness into a tube, tapering by stages towards its outer end. The upright was no doubt very light but cost, even in 1959, in excess of £50. What is more, should the car clip a kerb during racing – scarcely an unknown event – the upright would be bent and require replacement. The Standard one used by the 'amateur' cars, being designed for a passenger car of much greater weight, resisted such knocks and, in any case, if damaged cost only £1.50 over the local spares counter. When, in 1960, it was suggested to BRM that they might consider using the more rugged alternative, the reaction was 'What, and give up our beautiful machined all over part?'.

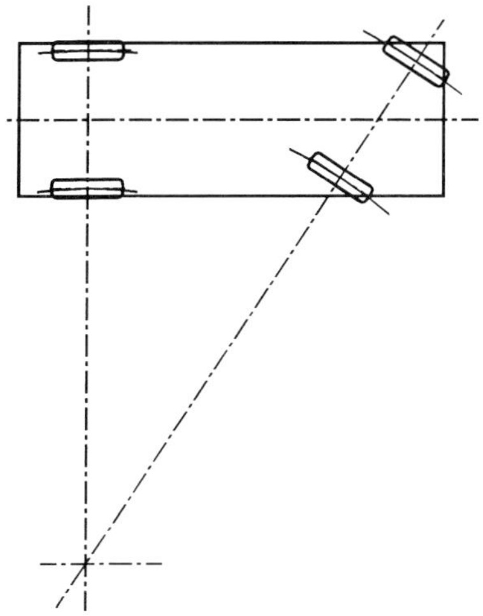

When the front wheels pivot as a pair (farm cart etc) the front and rear wheel centre lines always intersect at a single point, giving 'true' steering.

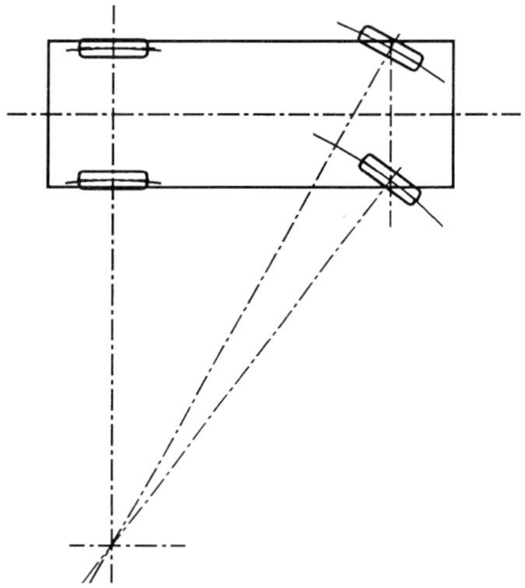

With separately steered wheels the linkage is designed to ensure the same single point arrangement. In doing this they toe out because the inner wheel turns more than the outer.

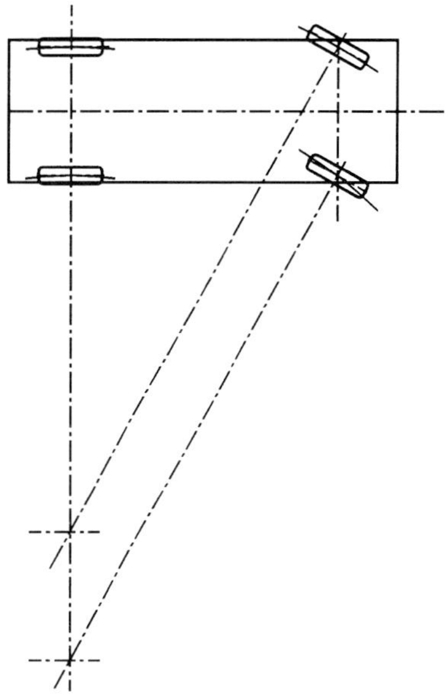

If the wheels have a parallel movement they effectively toe in as they are turned, because they both turn by the same amount.

Chapman was criticised over the rear suspension of the 18. Some said it had insufficient ground clearance. This was because of the purist approach to the suspension geometry. The forces acting on a car to make it follow a curve are generated at the point of contact of the tyre on the road. These forces act at an angle dependent on the cornering speed and corner radius. If the sideways force equals the weight of the car (a '1g' turn) the angle is 45 degrees. The forces in the suspension members will be minimised if they are as widely spaced as possible. Chapman therefore attempted to get the lower wishbone as low as could be so that it was as widely spaced from the drive shaft (the 'upper wishbone') as possible.

One result of this was to bring the wishbone fixings below the level of the wheel rim, it being only the tyre that kept it from touching the ground. If a tyre went flat the car would ride on the wishbone and to some this seemed wrong.

Subsequent rear suspensions which did not use the fixed length drive shaft as a member avoided the problem while allowing wide vertical separation of the members, since the positioning of the upper one was not dictated by drive axis considerations.

An indication of Chapman's grasp of what he was trying to achieve with a suspension design was given during the development of the Mk 20 Formula Junior car of 1960-61. During the car's first test day it was driven by a number of the regular drivers and by Chapman as well. The drivers were pleased with the feel of the car and had comfortably exceeded the speeds achieved in the previous model, the Mk 18. Chapman, on returning to Cheshunt, declared that it had not been fast enough through the corners and ought to have been three seconds a lap faster.

The springs used for this first outing were taken from a large stock of assorted springs that Lotus kept for experimentation. They had been selected to be as near to what Chapman had specified as were available. Chapman requested sideways deflection curves to be plotted for the front and rear

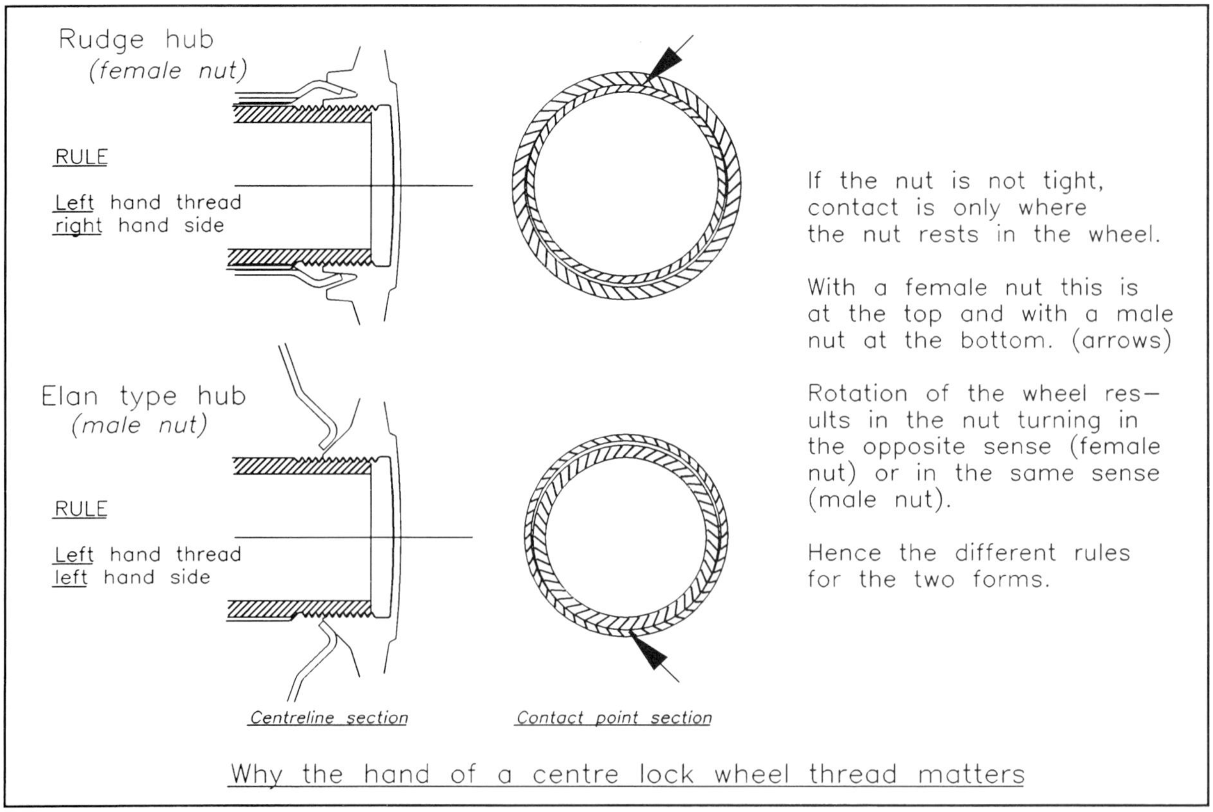

ABOVE **Knock-on wheel fixings. The centrelock wheel is held by a single nut which can enclose the centre of the wheel or enter the hole in it; whether the assembly rule is 'Left for Left' or 'Left for Right' will depend on which pattern is used. The drawing shows sections through two typical hubs, with the nut loose. In this condition the nut will be driven by the wheel as they rotate by so-called epicyclic action. It is the engineer's job to understand this and ensure that the wheel drives the nut so that it tightens.**

ABOVE RIGHT **Chapman helped his team to understand the effect by demonstrating with a roll of tape as shown. When the cap is moved with a clockwise circular movement (not rotated) within the tape, it represents the tubular part of the traditional 'Rudge' wheel. The tape (the nut) will turn clockwise as well.**

RIGHT **If it is the tape that is moved, it represents the 'Elan' style wheel resting on the nut. This time the nut, represented by the cap, rotates anti-clockwise.**

of the car. This was done by pulling it sideways one end at a time by means of a lever acting through a spring balance, while measuring the lateral deflection of the tyre with a dial gauge. By measuring the deflection at each increment of sideways load, a curve could be arrived at for the tyre in the actual situation on the car; it was done stationary, so was not perhaps a perfect analysis.

The figures were given to Chapman and the next day he returned with some new spring specifications. Springs were found from stock to suit and fitted to the car. At the next test day it exceeded its previous best by slightly more than three seconds a lap. No other changes had been made.

What Chapman had done was to recognise that his original spring choice was only as valid as the tyre characteristics he had assumed. These would have been from the maker's figures, it being the first time Lotus had used these thirteen inch sizes. Once 'real' curves were to hand, taking account of the actual weights of the car rather than those estimated in advance, he was able to re-calculate the frequencies required and so produce a better choice of spring. No doubt his learning from the work of Olly and Hodkin was brought into play.

An event which took place during the development of the 25 sheds a lot of light on several of the people involved. Carrying the concept of reducing unsprung weight to a minimum, Chapman wanted to be sure that the suspension parts were as light as practical. Sets were made up in different metal thicknesses with the idea that Jim Clark should test the car on each to find out the least that would be strong enough. This would have entailed driving until one set broke!

Wheel and tyre considerations

Closely allied to suspension design is that of the wheels. Wheels present some very particular design problems of their own, but in their role as part of the suspension the main considerations are lightness and stiffness.

In 1956, when Lotus first challenged the established Formula 1 contestants, all cars were relatively heavy and therefore hard on their tyres. Lotus (and certain other newcomers) were producing cars very much lighter and consequently obtaining greater tyre durability.

As a result of the relatively high rate of tyre wear it was necessary to change tyres during races (which were incidentally longer in those days). Knock-on type wheel fixings were therefore used almost exclusively. These in turn added more weight, again shortening tyre life.

Chapman reasoned that by turning to advantage all the weight saving he could achieve, he could design a car to run a whole grand prix without tyre change. This being so there was no need for quickly detachable wheels which brought the objective nearer by saving the weight of quick-change attachments.

The first Lotus design to put this reasoning into practice was the Mk 12 which used the famous 'Wobbly Web' wheel design for the first time. The wheel fixings were a complete throwback to the earliest origins of Lotus: the nuts used were from the pre-war Austin Seven. On some models further weight was saved by eliminating the hubs altogether. This was done by accommodating the bearings in the wheel itself. The 'wheel nuts' still visible from the outside were actually the brake disc fixing.

Magnesium wheels were not new in motor racing, having been seen on the Cooper and other 500 cc cars for some years, in the case of Cooper with integral brake drums. Such wheels were in style and engineering concept adapted versions of the so-called artillery wheel of the pre-war era, having a fairly flat centre and a spoked form. What Chapman brought was a new look at the wheel in using the principle of a constant thickness cast web – constant thickness castings are easier to make – arranged so that if flattened out it would be seen to form a (roughly) constant width strip of a length equal to the inner circumference of the rim. The strip was bent around the hub by folding the inner edge in a series of waves thus absorbing the spare length. Seen edge-on, the web is wide at the centre where it attaches to the hub, tapering progressively to its natural thickness at its outside where it meets the rim. This is in many ways an ideal form since the main forces on a wheel are the cornering forces which tend to bend the web of the wheel and are at a maximum near the hub. The wobbly web thus places the material efficiently to withstand the loading. The edge-on view of the design has a similar form to that of the classic wire wheel.

The continuous web, because it bridges between the much greater stiffness of the rim and of the hub, is stabilized at its edges so that it cannot deform in the directions in which it is weak of itself.

Later wobbly web designs, such as that used on the 20 FJ car, carried the idea further by incorporating sections of almost spherical shape, thus importing something of the stiffness of the ideal monocoque, the eggshell.

As wheel requirements evolved and they became wider in relation to their diameter the set of conditions which spawned the wobbly web wheel, often called 'wibbly wob' in the Lotus drawing office, gradually changed until the design was abandoned in favour of (again) a spoked form. It is interesting to recall that a rear Formula 1 wheel of the 1950s would be typically 15 or 16 inches in diameter and only 5½ wide. Compare this with 13 diameter and 16 inches widths of the 1980s.

Chapman did not abandon the use of centre lock wheels for all designs, since they clearly had a place on road cars or in racing cars where race distance necessitated tyre changing. Road cars up to the Elite of 1957 had wire wheels which were used partly because they were 'normal' on any car with sporting intentions. By this time pressed steel wheels of ordinary saloon cars were lighter and more rigid than wire wheels. Chapman therefore combined these benefits with the centre lock principles used on wire wheels when designing those for the original Elan.

The standard hub design for sports cars since before the second war had been the splined Rudge hub, made in several sizes according to needs. The system used a pressed steel hub member which formed the centre of the wire wheel for which it was conceived. The simplest way of adapting the idea to an all pressed steel wheel would have been to form such a centre in a normal wheel centre pressing. This would mean a lengthy tubular section added to the wheel which would add weight and manufacturing difficulties.

Chapman's solution was to raise a conical section in the middle of the wheel disk, reversing the form at its extremity to accept a nut with a corresponding male cone. This was really a larger version of the embossing used to provide each of the nut seatings on a bolt-on wheel.

Drive between the hub and the wheel was to be via five pegs fixed where the wheel studs would be on a conventional hub.

Meeting the design needs this way meant that the well known rule 'right hand thread on the left hand side, left hand thread on the right hand side' of the Rudge system had to be reversed. The result, though perhaps easier to remember, was sure to cause some confusion. Indeed, it is known that at least one Ford GT40 (which used the same male nut form) suffered in the 1970s from extreme wheel loosening problems because the hubs had been refitted on the wrong sides following a rebuild.

To start with, even among Chapman's design team this phenomenon was not understood and he therefore gave a demonstration of the principles involved using ordinary office materials. The mechanism that keeps the wheel nuts tight on both designs is the same and is the principle of the epicyclic gear, although there are no teeth. With the female nut of the Rudge design the car 'rests' on the upper half of the nut's circumference. Changing to a male cone means that it rests on the lower half, thus reversing the epicyclic drive direction.

6 Of Chassis Tubes

To THE INEXPERT eye, different makers' multi-tube space frames used prior to the monocoque concept were of similar appearance, only differing in the number of tubes they used.

To the engineer, however, the differences are much more fundamental. Chassis designed along a true space-frame approach make full use of triangulation principles so as to get the maximum strength and (more importantly) rigidity with the minimum of weight. The way a triangulated structure gets its strength by avoiding bending stresses in the material was explored in an earlier chapter as far as a two dimensional structure is concerned. Evidently the same ideas are applicable in a three dimensional structure like a chassis.

A disadvantage of space frames is that they demand straight tubes; and large unobstructed spaces between tubes can only be provided with difficulty – despite the name.

For this reason designers have often chosen to compromise by making use of curved tubes, or tubes theoretically wrongly placed, so as to allow more access to the inside of the structure at the expense of weight or stiffness. An ideal space frame will not only employ the tubular members in tension or compression, but will also have an absolute minimum of them. Chapman made a speciality of aiming for 'perfect' structures and the example of the Mk 17 chassis already cited is a case in point, although it was designed for him. Later models normally had to sacrifice the second of these ideals in order to give reasonable access to the mechanics of the car – to say nothing of getting the driver in.

Since every rectangular tube pattern within a structure must have a diagonal member to prevent 'lozenging', a particular difficulty arises when the rectangular area is under the driver's seat, since to insert a diagonal tube will mean that the seat cannot be mounted so low down. The lower the driver sits the lower can be the centre of gravity of the car, which brings advantages in road holding as well as reduced air drag.

To avoid this problem Chapman used a metal sheet to fill in this particular place on the early rear-

(Early 1960) A type 18 Formula One car is seen in the sheet metal shop of the then new Cheshunt factory, ready for its aluminium parts to be fitted. The highly-shaped seat-back fuel tank can be seen. It is not only shaped to fit in the space behind the seat but also has a triangular extension protruding between the chassis tubes. On later models (21, 24) this shaping became quite extreme and the tanks were difficult to make and harder still to render entirely leak-proof. Ian Lee-Smith, an engineer with Lotus at the time, stands behind the car.

ABOVE **At the 1960/61 Racing Car Show Lotus unveiled the Mk 20. This car replaced the 18 and was much more compact and of a notably elegant and streamlined shape. Colin Chapman looks pleased with the impression it has made as he poses for the picture taken for *Autosport*'s show report. The appeal of the new shape can be assessed from the fact that it continued to reappear almost unchanged in many subsequent models including the 21 (F1), 22 (FJ), 24 & 25 (F1) 27 (FJ), 29 (Indy), and 33 (F1).**

THOUGHTS ON THE LOTUS 20 . . . !

ABOVE **Such was the incredible 'smallness' of the new car that this cartoon appeared in *Autosport* the week following the show report mentioned above.**

David Piper bought the Lotus 20 prototype after the Racing Car Show as it stood. Chapman allowed it to be sold as it was even though he knew of the deceptive mods it had for the show. Piper raced the car in Italy and France and at one town circuit had the misfortune to hit a wall with a rear wheel, breaking the suspension upright. Rather than forego his start money for the next race, Piper strapped the upright with steel strips held on by wood screws – all he could get hold of. He later took the car back to Cheshunt to buy spares and to ask Chapman for help on the non-standard parts. Chapman was unwilling, even though the rear tyres had been worn away on the suspension as a result of the wheel modification. He handed Piper over to the author with the instruction 'get rid of him'. Because it seemed that Piper was gallantly 'waving the flag' on a shoestring, the author was able to 'find' the necessary spares for Piper at an advantageous price.

RIGHT **When Chapman was starting to compete regularly in Formula One races, people were concerned that the lighter weight of his designs might constitute a risk to the drivers. Chapman wrote this article to give a counter argument, for *Motor Racing Magazine*, October 1958.**

COLIN CHAPMAN EXPLAINS WHY

LIGHTWEIGHT CARS ARE SAFER

A GREAT deal of publicity has been given to Juan Fangio's alleged statement that racing cars are becoming too light—and therefore too dangerous—but most of the people closely connected with motor racing, in this country at least, seem to be in agreement that this is not the case.

Two fatal accidents in the Grandes Epreuves, which previously had a wonderful safety record, have been the subject of a great deal of unfortunate—and often ill-informed—publicity; although perhaps purely coincidental, it should be pointed out that these accidents involved the heaviest of the current Grand Prix contenders. Thus, although it would be unwise to say that the "minimum weight limit" addicts—and there are quite a few of them, even in Britain—have got hold of the wrong end of the stick, it would seem to be very easy to show that there is a very strong argument on the other side. It must be obvious that the heavier the car the more difficult it is to control at the limit of adhesion and to slow it from high speed (due to weight increasing its kinetic energy) while in the event of impact with "immovable objects" the extra inertia of the heavier car is likely to result in more damage to everything, and everybody, concerned.

Fangio's statement—if he ever made it—was publicised because Fangio is the acknowledged World Champion driver, and his experience could continue to play an important part in motor racing affairs long after he has finally retired from the cockpit. Moss, whose name is synonymous with fast driving to the general public in Britain, and indeed throughout the western world, deserves equal attention on this subject, and he has said that, given adequate roadholding, the lighter car has the advantage in every way.

Thus, as far as the designer is concerned roadholding is the critical factor. It is, of course, a subject worthy of a complete study on its own.

The roadholding of a Formula 1 racing car is obviously far superior to that of most road cars but this alone demands a far greater awareness of the eventual limits of adhesion on the part of the driver. A close analogy can be drawn between driving and flying, for in the latter context there is a critical angle (related to speed) at which the wing of an aeroplane will lose all its "lift," and consequently will stall; much the same is true of tyre adhesion on a racing car. Whatever the design of the aerofoil section there is a point at which any aeroplane will stall. In the same way, whatever the design of the suspension there is a definite limit to a car's roadholding, and this is largely determined by the tread pattern of the tyre—and, of course, the condition of the road. As the pilot can judge the stability of his plane by correlation of speed and flying altitude so the driver should be able to balance his car against the forces which are tending to send it off the road; the best drivers have this faculty in the seats of their pants.

The maximum slip angle which the tyres can adopt without losing all adhesion is thus the criterion, but the chassis designer can delay the point at which all "grip" is lost by endeavouring to keep as much of each tyre as possible in firm contact with the road, and by equalising weight distribution.

To achieve these ends, it is now generally agreed, the designer needs to use a space frame chassis—the space frame being synony-

Allison's car at Oporto, after all the pieces had been collected together.

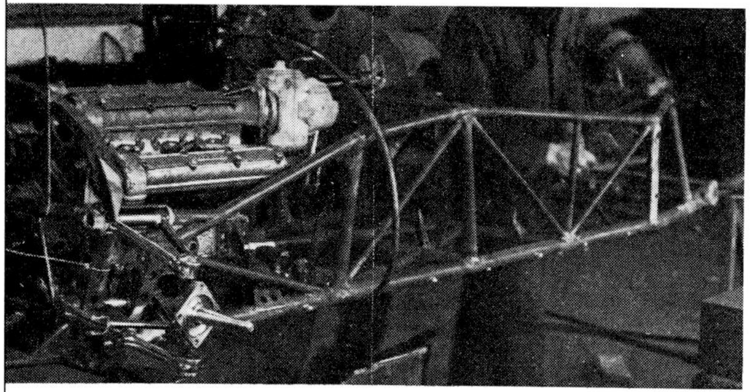

Inefficiency of the Ferrari F1 space frame results from the use of unequal diameter tubes, the upper tubes taking more load than they should. This frame also has a minimum of diagonal bracing.

mous with light weight and structural efficiency—in conjunction with some form of independent suspension to all four wheels. On this latter subject, of course, there is not yet complete unanimity of agreement, or rather agreement in principle has not yet resulted in universal adoption, in practice, of independent rear suspension. The de Dion system—still widely used—can have many advantages; fixed camber angle under roll; no rear wheel steering due to suspension travel; constant height of roll centre. The disadvantages can be high unsprung weight, the need for sliding splines in the transmission system and—by way of anomaly—constant height of roll centre; this latter can have a very definite influence on cars with variable weight distribution, that is those with fuel tanks at the rear, which for real efficiency demand a self-compensating suspension system, such as on the rear end of the current single-seater Lotus.

The basis of an efficient racing car, however, is the frame, the layout and stressing of which must be planned with reference to the function of each and every tube. Indeed, until the final diagonal member is welded into position a frame of tremendous rigidity may have no torsional strength whatsover. But once completed such a frame will never fail unless subject to outside influences—that is, if the driver hits something.

IMPACT ABSORBED

And this is where we return to the safety factor. As an example we can take Cliff Allison's practice accident at Oporto. On a left-hand bend Cliff got into a slide (to the right), overcorrected, slid to the left and—at very considerable speed—the rear end of the Lotus struck a lamp post. The rear suspension was torn off—plucked off might cover it more adequately—and in a situation in which a heavier and more "robust" car would almost certainly have slewed round and rolled over the Lotus remained on an even keel, lost its front suspension to the next lamp post, and came to rest looking very much "second hand." The great point was that Cliff got out completely unhurt.

Examined slightly more scientifically, this accident had no ill effect on the driver (other than on his confidence) because firstly the lightweight car dissipated much of its kinetic energy in sliding across the road, in fact between the point at which the car was committed to an accident and the actual point of contact. In the collision with the lamp posts the extremities of the frame took the shock of impact and absorbed it in depth; the driver was not subject to a rate of deceleration at which he would be almost certain to sustain injury against parts of the car. The suspension units, which have been shown to be strong enough for all normal use, did not provide sufficient resistance to throw the car into the air and overturn it, but rather aided the frame in absorbing shock. And although the extremities of the car were very much beyond repair (the engine also was damaged, and the radiator was found some distance away) that part of the frame which surrounds the driving compartment remained intact, so that Allison was able to climb out in the normal way as soon as the car came to rest; an expensive accident as far as the car was concerned, but at least Cliff was able to start in the race—even though it was only in another car.

Weight-saving is a process which is naturally open to criticism if it results in unreliability, or breakages under normal use, but there is no doubt that a well-designed and properly constructed lightweight car can stand up to very hard use, the mechanical parts (engine and transmission) failing far more often than the chassis components. In this connection it must be emphasised that design is more important than material; despite what might have been said about exhaust pipes dropping off our Formula 2 car it was the design—not the gauge of the tubing—which caused fracture, and it doesn't happen any more. In the same way the original Formula 2 car suffered frame failures when the rear suspension was changed from de Dion to strut; the different load paths required for this layout have since been incorporated.

Instead of reducing weight, however, it is better to design light in the first place, modifying any components which may show signs of stress during development work. This, in the case of the frame, means having links which are all of the same strength, and in this connection it can be said that the Ferrari, although it wins a lot of races (thanks to the most powerful 2½-litre engine in current use) has a frame which is inefficient.

The total weight of a racing car on the starting grid, with driver and fuel aboard, is in the limit determined by the weight of the engine; if this can be decreased by one pound then the rest of the car can be made four pounds lighter. In all this, however, tyres and wheels are the critical factor and their weight has a considerable influence on the car as a whole. The job of the designer is to keep the right relationship between sprung and unsprung weight, and, above all, to keep the greatest possible amount of tyre tread in contact with the road.

So in the end it all comes back to roadholding, and given the suspension (and tyre) qualities which enable the driver to go round corners at very high speed without losing adhesion, the lightweight car will not only have the advantage of safety—in the event of the driver "over-cooking it"—but will also be in a position to defeat cars with more powerful engines and perhaps greater maximum speed. And that is what the Formula 1 Lotus is going to do next year.

A Lotus Eleven space-frame. Plan stiffness for the frame is achieved by the close riveting of a strong alloy sheet to the underside; this also forms the floor and undertray of the car.

OCTOBER, 1958

ABOVE **A view of the chassis of the Formula Junior Mk20. This car was the first prototype shown at the Racing Car Show of 1960/61. Note the internal exhaust system, later abandoned; the 'show' springs are fitted. These were selected to mislead rivals at the show who were known to measure up details like these to copy. The rear springs are at more than twice the correct strength and those at the front only half what they should be. The dampers are doctored to hold the car at the right height. The rear wheels also had special treatment; they had been cut back so that the track was greatly reduced. The car was later bought by David Piper and raced extensively on the Continent.**

LEFT **A head-on view of the 20. The Lotus badge and the 13 inch steering wheel give an idea of scale. In 1961 this ground clearance was considered minimal.**

engine single seaters. On the Lotus 18 the underside of the seat touched the sheet metal reinforcement of the chassis. The seat back which sloped at 60 degrees, dictated the frontal area of the car by the posture of the driver. The ground clearance of the car was about four inches and so the whole size of the car was a result of a few parameters – starting from the angle of the seat back.

For the 20, which succeeded the 18, the frontal area was reduced by lowering the seat back angle to 45 degrees so the driver was more reclined. One of the unexpected results of this was that when the drivers, who knew the racing circuits from racing on them in 18s, came to drive the 20 they found that there were horizons appearing where the road had appeared flat before. This was just because the eye level had been lowered by three inches or so. When the 20 came out it surprised all by its extreme low build. Such was the impact of the approach. Its compactness, its appearance, in short its 'rightness', now makes it a landmark in the design progress of Lotus single seaters. The same basic shape can be recognised through several succeeding models in the different formulae.

ABOVE **The 20 seen alongside its predecessor the 18. This side view brings out the size difference dramatically.**

BELOW **The view head-on emphasises the height difference. Because the driver sat lower false horizons appeared on stretches of circuit which had clear visibility in an 18.**

ABOVE **An early prototype Mk 20 on a day's testing at Goodwood in February 1961. Colin Chapman discusses the work with the team. The author is second from the right and at the left is Peter Warr, later team manager and now one of the elders of Formula 1. Mike Costin is hidden behind the mechanic near the rear of the car.**

LEFT **In this view the car is seen with Mike Costin on the right, Mike Taylor (Trevor's brother), bending, and the author. Note that this car has a Renault gearbox in place of the VW unit in the first car, and that a heat shield has been added under the exhaust pipe to protect the brakes – this was to prove inadequate.**

In designing tubular chassis Chapman adopted a very pragmatic approach, as indeed he did with most things. This is to say that in important ways, such as in the triangulation of structures, he followed purist principles, whereas in certain respects he deviated from the theoretically perfect in the interests of practicality. The use of mild steel tube for chassis and suspension parts is a good example of this.

Chapman always used this cheaper and more forgiving type of material where other designers would go for high strength steels which would permit a lighter structure to be designed. The drawbacks to these in Chapman's view included the fact that they required much more careful handling in welding and during manufacture. This also meant that they were much more difficult when it came to making hasty repairs – repairs which racing cars often need. He did, in fact, specify a high strength form of welding which resulted in much higher strength than conventional fusion welds. This was nickel bronze welding, much used in aircraft construction at the time.

The advantage of this form of weld is that it is made at a relatively low temperature and therefore does not soften the steel of the chassis components. Remember that, although the steel used is described as mild, it is not in fact soft, since the processes involved in forming it into sheets and tubes hardens it greatly. Normal welding, in which the material is heated to melting point, softens the steel for some little distance around the weld, thus the whole structure is weaker. Low temperature welding preserves the extra strength in the mild steel and so uses it efficiently.

Another way, although related, in which Chapman turned away from the theoretical towards the practical was in the question of stress relieving.

The last tubular chassis Formula 1 car was the 24, very much in the shape of the time. The 25, so as to reduce the number of new features to be developed, had everything other than the chassis as exactly identical to this model as possible. Trevor Taylor is seen on his way to second place in the Dutch Grand Prix in May 1962.

The 27 Formula Junior was the first Lotus in this class with monocoque construction. Chapman hoped that with the lighter weight the full '25 style' structure would not be needed. The first 27 was built of an aluminium folded channel in which the driver and drive train were located, with a glass fibre moulded outer panel each side to provide the box stiffness. In practice this proved too flexible and later cars were of metal structure.

When a structure is welded the material is left with internal stresses due to the concentration of heat at different places at different times. In effect, the different parts of the structure are 'fighting' against each other – again lowering the overall strength and causing some distortion. The approved treatment is to heat the whole structure to a high enough temperature to allow it to stabilize without the risk of softening. Lotus chassis were not normally stress-relieved in this way because the real gains were not felt to justify the cost and effort involved. In any case an emergency repair could not be treated this way, but had to survive.

One of the results of not stress-relieving caused some inconvenience. Most of the chassis had removable sections for engine access. Such sections also suffered distortion due to the residual stresses which, of course, never matched that of the chassis for which they were made. This meant that they were difficult to fit and had to be levered into place when they were refitted.

On one occasion a Lotus Formula 1 tubular chassis was stress relieved. The chassis was that of a car being driven by Innes Ireland at Spa. The car caught fire and was more or less destroyed. The magnesium wheels burnt and there was a river of aluminium down the road from the engine! The reaction of the team was to scrap it there and then, but they were obliged to bring back the remains as otherwise they would have been deemed to have exported it and duty would have been payable.

When the wreck was back at Cheshunt it was found that with cleaning and a little work the chassis could be re-used. The removable parts of this chassis fitted perfectly without the usual straining due to the stress-relieving action of the fire.

Lotus tubular chassis of the early 1960s were jig built to finished dimensions. The suspension mounting points consisted of machined bushes or plates welded into place in the chassis, with no subsequent machining. (The low-temperature welding made this possible, since there was no severe scaling of the parts as would have happened with fusion welding.) Other makers of the period went in for the purist approach to chassis and suspension component design.

BRM, for example, used high strength Reynolds 531 tubing for the chassis structure, and their suspension parts were a work of art in themselves. The chassis were built in jigs and the suspension and other fixing points jig-bored following stress relief and heat treatment of the whole chassis.

The suspension parts were, perhaps, exemplified by the rear radius rods. These were made up from thin sheet steel in two pieces. Each was cut to a carefully calculated profile and rolled into a tapered tube. The tapering was not straight but followed a mathematical curve to achieve maximum theoretical strength for the weight. The two pieces were then welded together by their large ends to form a tube with the maximum cross section at its centre, to coincide with the area of maximum

BELOW **Another 20 & 25 descendant was the 29. This was Lotus' first Indianapolis car and only the second monocoque design completed. Len Terry was the designer for this model. Most of the work was done in a tiny office above the racing shop shared with the author and Brian Luff.**

ABOVE RIGHT **Also exhibiting ancestry from the 20 is the 34 Indianapolis car, which achieved pole position there in 1964. Clark is seen racing it at Trenton.**

BELOW RIGHT **The type 33 might be called the definitive or developed 25. The shape inherited from the 20 is clear, although by now showing the slimming which was to progress further yet.**

6

52
CO-83
TEAM LOTUS
1

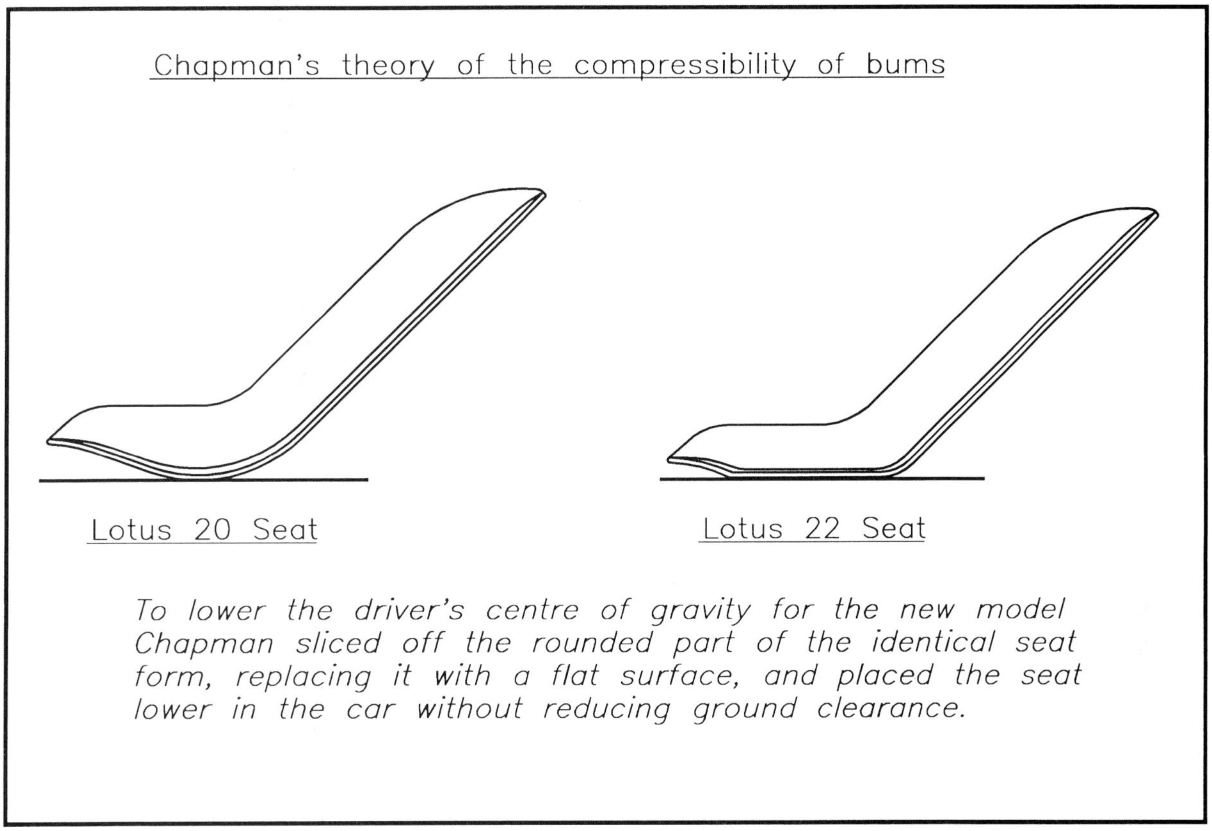

ABOVE LEFT **At the 1963 Racing Car Show Lotus 22 and 23 are seen side by side. These two models were very much designed together. They share all mechanical parts where possible. Apart from the chassis and bodywork and a few lesser parts, only the rear radius arms differ, due to the wider frame of the two seater. The author was the Project Engineer for these two models and prepared this pair for the show. The 22 had all mechanical parts, including wheels and engine and gearbox in black stove enamel, with chrome plated details, and the body super finished in white. The 23 was the opposite, with the black body contrasting with white enamelled wheels, engine, gearbox and other parts. The author was later able to buy the show-finished black wheels of the 22 for his 7/20, shown earlier. The front brake assemblies from the nearly completed 7/20 were lent to Lotus for use on the first 23, as their new installations were not then ready.**

LEFT **The same type 22 is seen at a circuit test day. Ray Parsons is talking to Peter Arundell in the car.**

ABOVE **For the Lotus 22, which replaced the 20 the following year, Chapman again wanted to lower the driver. To have chosen a more reclining posture would have forced an increase in the wheelbase and length of the car, which was not acceptable. With ground clearance down to three inches there did not seem, much room for manoeuvre until he came up with his 'Theory of the Compressibility of Bums'. The seat form was a smooth moulded shape in glass fibre, nicely rounded under the driver's bottom. Chapman realised that if a slice was taken off the seat form at its lowest point and replaced with a flat surface, the seat could be mounted lower in the car and the driver would sit lower without having to be more reclining. This gain was made without drastic changes in any other part of the car.**

bending load. The outer ends were flattened and welded to rough machined rings in a jig, reinforcing plates being wrapped over them. The rods were then heat treated, first to remove any stresses remaining after welding, then to develop the strength of the material (which had been additionally softened by the extensive fusion welding). Next the end fittings were jig-machined to receive the ball joints, and last the whole assembly was shot peened. Shot peening is a process of blasting the part with small round particles (like shot) which compact the surface all over the part to make a sort of skin. The fatigue resistance can be considerably increased by this process.

This procedure is not described at length here because it was somehow inferior. On the contrary, it was in line with the best known technology for producing the strongest possible parts from a minimum of material.

Chapman's approach for the same suspension rod was to take a length of plain mild steel tube of parallel section and nickel bronze weld to each end a pre-machined bush, made to receive the ball joint. No heat treatment was involved and apart from chrome plating or other finish, the part was ready to use straight from the jig.

The Lotus part was undoubtedly heavier than it could have been but the balance of priorities for Chapman was such that the low cost, the speed of manufacture and the ease of repair compensated for all practical purposes for the extra weight.

Chapman would reason that on a part weighing only a few ounces, only a few ounces can possibly be saved, therefore go for weight savings on the larger parts of the car.

While dealing with the subject of tubular chassis designs it is worth looking at the question of straight versus curved tubes. As has been explained earlier, most materials stand stress worst in bending. A curved tube subject to a load will inevitably suffer a bending stress due to its shape. It is in principle less efficient than a straight tube doing the same job. This is the theoretician's view, but the practical designer will use curved tubes judiciously when, for example, straight tubing cannot be made to suit the particular location. It may be that a tube has to be 'dodged' round part of the engine and in such a case it could be that a curved tube is a lesser

LEFT **The design of the Lotus 23, from what might be called the mature period of tubular frames, can be used to point out some typical Lotus features. First, note the rear suspension top bolt. It serves as a mounting for the suspension top link, the spring/damper unit and the anti-roll bar. It is inserted from the front so that the anti-roll bar can be changed easily without disturbing the other parts. Next, the engine bay bracing: there is a linkage of tubes that can be traced, via the removable centre section, from the body supports ahead of the rear wheel at one side, behind the engine and over to the opposite body mounting. The main chassis tubes are used to carry water and oil, and the water temperature sensor can be seen in the right hand tube close to the instrument panel. A very simple handbrake is provided by operating the brake pedal itself from a small lever (fixed near the water temperature sensor). The movement is transmitted through a flexible cable which can be seen just behind the radiator. Notice that the rear suspension uprights are made identical left and right. Lugs are provided which on the front side serve as attachments for the anti-roll bar links while acting as brake caliper mountings at the rear. From the windscreen the form is followed round to the door 'windows' which are wrapped right over past the horizontal to blend with the rear body.**

evil than the extra structure necessary to produce a pure result.

Many designers have gone much further than this minimal use of tubes subject to bending and produced complete chassis structures with no attempt at eliminating bending loads. One of the objects of such designs is to avoid places where several tubes are joined at one spot. This is because such places can give rise to additional distortions due to difficulties in controlling the welding temperature. Chapman's view of the welding problem was that it had to be faced as the alternative was too horrible to contemplate.

It would seem that another attraction of such designs is that the frame can be designed in isolation and the suspension fixings added later. The Lotus approach was to design the car first, starting from the overall package of engine, driver, axles. Next would come the suspension geometry layouts incorporating the lessons learned since the last time. These would give rise to a knowledge of where the suspension mountings needed to be. Not until then would the chassis design start. The main tube positions would be decided to minimise the offset of the suspension brackets and get best lines of action of the forces involved. So the chassis would grow progressively into a complete concept, matching as nearly as possible the ideals of design theory as well as the practical requirements. It would be created after the loads to which it would be subject were known as closely as possible. If any serious flaws were found in the suspension geometry it would be practically impossible to make changes to suspension mounting points.

The Cooper approach clearly did not have quite this limitation, indeed it could be a positive advantage in manufacture. The Cooper chassis had its suspension brackets finally positioned only after the welding distortion had taken place on release of the chassis from the jig. Thus some compensation was made for this in their positioning. Chapman, of course, would argue that it is better to have an 'ideal' chassis despite the advantages of Cooper's approach.

Front end of a Lotus 33. The ends of the steering rack positioned to align with the pivots of the top arms, and the angled steering arms to bring the outer ends of the track rods in alignment with the top arm outer ends. The oil tank behind the radiator is cut away to allow the flow of air. The oval shape of the radiator is flattened to provide for the slim ducts taking cooling air to the suspension units. (Cf p.84.)

All the competitors quickly brought out monocoque cars to match the 25 – by which time Lotus had the 33. Most of them adopted many of its good ideas (or perhaps already had them, for they were not all new) and the degree of imitation varied. This shot shows a BRM.

7 Monocoque

THE MONOCOQUE CONSTRUCTION – following the principle of the eggshell – is of advantage to the designer seeking light weight and high strength. Chapman, like all engineers, had a sound appreciation of this from his training. His first attempt at a monocoque chassis concept, though, was not confined to a simplified experiment but was a full road-going closed car in the grand touring idiom. The Lotus Elite was at the same time the most sophisticated car he had ever conceived, the first car he had attempted with full sports saloon levels of trim and comfort, the first production road car to make use of a version of the Coventry Climax engine and also the first car from Chapman's brain intended for the non-enthusiast production road car market. Previous models were really for something other than daily use, although they were put to it by enthusiasts.

It was a most immense undertaking for any team, bearing in mind the number of 'firsts' it involved. It could be said that no fully established motor manufacturer would have taken on such a task, even with resources a hundred times greater than had Chapman had at his disposal. This is because such companies were very much aware of the risks. Most manufacturers undertake only modest technological steps, but to have tackled so much that was new and unfamiliar at the same time, and with such a small team, was typical of Chapman at that stage. While looking at the 'firsts' notched up by the car it must be remembered that it was not only Chapman's but the world's first car with a fully glass fibre structure.

The concept of the car was that it should be very low and light, be highly refined from an aerodynamic point of view and possess the handling and performance for which Lotus was coming to be seen as master. The car was to have the looks to compliment those of the racing Lotus Elevens which were dominant in their class at the time.

Looking back now it seems that a team of around twenty, although working with Chapman's guidance, could not possibly have tackled such a task. But such is the stuff of pioneers.

Structurally the car was an imperfect monocoque. If it had been made perfect it would have had no openings and wouldn't have been a car, but an egg. The egg is a perfect use-once-only monocoque. To provide the wide doors, the window apertures and accommodate the mechanical parts of the car necessitated careful planning in order to maintain the natural properties of the true monocoque.

One of the requirements that Chapman demanded of the designers working with him was that no unfinished glass fibre surfaces should be left exposed in the structure.

Glass fibre mouldings were to be used, made by hand. Such mouldings have one smooth side from contact with the mould and one rough side where the layers of glass and resin are built up. To avoid any trace of this rough side being visible meant that all the sub-mouldings should be bonded together so that the unfinished surfaces formed the inside of a closed unit. While such a policy was compatible with the strength and weight aims, it set considerably more problems in achieving it.

When a structure is being designed using a stressed skin or monocoque principle, extra strength has to be added around any openings needed. This is because the strength comes from the continuity of the material, each section supporting and being supported by its neighbours. Cutting out for a windscreen is therefore in conflict with the stiffness requirement. It is easy to see that, if the roof is to fulfil its role as a major bridge piece from front to back of the body, it cannot do so if it is only attached by what is left when the windscreen hole has been cut out.

On the Elite a steel tube hoop was built in from the undertray level at one side of the body up by the door hinge area, up the windscreen pillars and round the top on the screen, and down to the other side at the undertray. At each of its ends the tube was made to protrude slightly, so serving as a hard point by which the finished car could be jacked. This approach of using one part for many functions is typical of Chapman's philosophy and, to be fair, an approach used by many talented engineers.

Box forms are very popular with engineers for their high strength to weight ratio. In the Elite there were eight such box sections. Two of these were in the areas behind the front wheels, where they provided the main stiffening between the mid-car area and the front suspension, also providing support for the engine unit.

For a long time Chapman had been concerned with how to carry fuel in a car for the minimum cost in terms of weight. He had proposed (though perhaps never tried) an idea for the single seat racing cars of making use of the tail body work. The scheme was to spray round inside this section with a petrol-resistant compound to seal the rivet holes etc. Had this idea been workable, the fuel could have been contained in a part of the structure that was there anyway for very little added weight.

When it came to the Elite, an idea like this was used in making sure the two large front box sections were carefully sealed, providing them with filler caps, gauges and feed pipes and thus creating almost 'weightless' fuel tanks. These tanks each held around 25 gallons and added in the region of 15 pounds of weight. This compares very favourably with the traditional figure of 1 pound per gallon used as a yardstick for conventionally constructed fuel tanks.

Space was saved by the incorporation of the fuel tank with the front wings in two ways, since both

The Elite set new standards for elegant simplicity of line in sports cars. This one is the show car. Being such an early prototype, the structure was not all glass fibre as intended. Much of the under structure was in aluminium. Albert Adams was required to coat the panels with glass fibre lay-up so no one would know.

the passenger compartment and the boot space could be maximised. This idea, hailed as brilliant when first seen, was in fact dropped later for several practical reasons. First the outer skin of the car was also the outer skin of the fuel tank. Thus any damage or cracking in use in that area of the car could cause a fuel leak. Secondly people love putting badges on the sides of their cars, and these often necessitate drilling holes for rivets . . .

One way in which glass fibre construction is better for the designer than metals is that its thickness can be varied in a progressive and calculated way in proportion to the loads it is expected to carry. With metals this is not practical with conventional and economic methods of body manufacture. In the design of the Elite, much was made of this property in reducing weight. Of course, this meant that the performance of the finished body was critically dependant on the correct execution in manufacture of the local thicknesses determined.

The prototypes were naturally built with great loving care by people who knew what was required. However, once the car went into series production it had to be entrusted to outside contractors. The first of these, the boat manufacturer Maximar, had in any case been involved in some of the early stages and were perfectly able to reproduce the structure as needed. But the Bristol Aeroplane Company, a subsequent supplier adopted to meet increasing volume requirements, was unable to maintain control initially and the first bodies Bristol made were not strong enough in the areas of attachment of the rear suspension.

By this time production was in full swing at Cheshunt and the stock of finished and painted bodies was stored in the open behind the factory. When the weaknesses in the Bristol bodies came to light (and it was the customers who found them) something had to be done to rectify the stock bodies. For this reason a re-work was designed and Bristol was obliged to send a team to cut open every body and reinforce it in the agreed way. For weeks there were caravans in the Lotus employees' car park in which the team lodged while they worked their way through the stock.

Through this experience another lesson was learned. Concealing the rough side of the glass fibre meant that no areas were visible for inspection, so that completed bodies could not be checked. It was for this reason that, although a percentage of the stock shells were perfectly correctly made, inevitably they all had to be opened up.

It may be that this problem stemmed from the fact that Bristol was a much bigger, longer established and more highly proceduralised concern than was Lotus at that time. The company worked effectively from detailed written instructions, drawings etc., but could not ensure that verbal explanations were adhered to. Lotus, still to an extent a bunch of committed enthusiasts, did not have available the complete set of written specifications necessary for the Bristol systems.

Years later, when Lotus was looking to produce the Elan, Chapman proposed another chassis-less glass fibre structure. In its initial conception the car was to be a 1,000 cc open car, a Lotus equivalent for the Sprite and Spitfire. The customer was seen as someone in the market for a car of that group but able to pay more for the extra performance and road holding Lotus knew they could offer. The engine was to be a double overhead cam version of the Ford engine, which was at that time known only in the Anglia 997 cc version. The original (reduced size) clay styling model of the Elan was made on this assumption. When it became clear that the Ford engine would become available in other versions (starting with that for the 1340 cc Classic), Chapman and his team decided to go for the greater potential offered if the car could accommodate the larger sizes. The clay model was accordingly modified by raising the bonnet line ($\frac{5}{8}$ inch) to

This close-up of the Elite suspension serves to show the clean underside of the car. Also the fact that there is no clear dividing line between the 'side' of the car and the 'bottom'. This caused difficulty with owners who could not tell where to stop washing. On the Elan a definite corner set the limit.

ABOVE An early Elite. This famous car was the first in the hands of a private owner. It belonged to Ian Walker who is seen in it at Silverstone. The car was first roadworthy at 3.30 am on 10th May 1958 and by 10 the same morning was practising for a race at Silverstone that afternoon, which it won. It won again the following day at Mallory Park. The shape of the car was very pretty and it has become a classic of its time. Note that the side windows slope in so much that it is impractical to provide for a wind-down system. They were made to open slightly at the rear for ventilation, or to lift out entirely and store inside the car. Being of Perspex, they were light and flexible. The aerodynamics were such that at speed with the windows removed the driver was only faintly aware they were missing, there being no violent inrush of air.

BELOW This rear view of Ian Walker's Elite shows the neat rear end and the stainless steel bumper. Inside the rear window can be seen the top strut mountings. On later cars these were smaller and smoothly rounded.

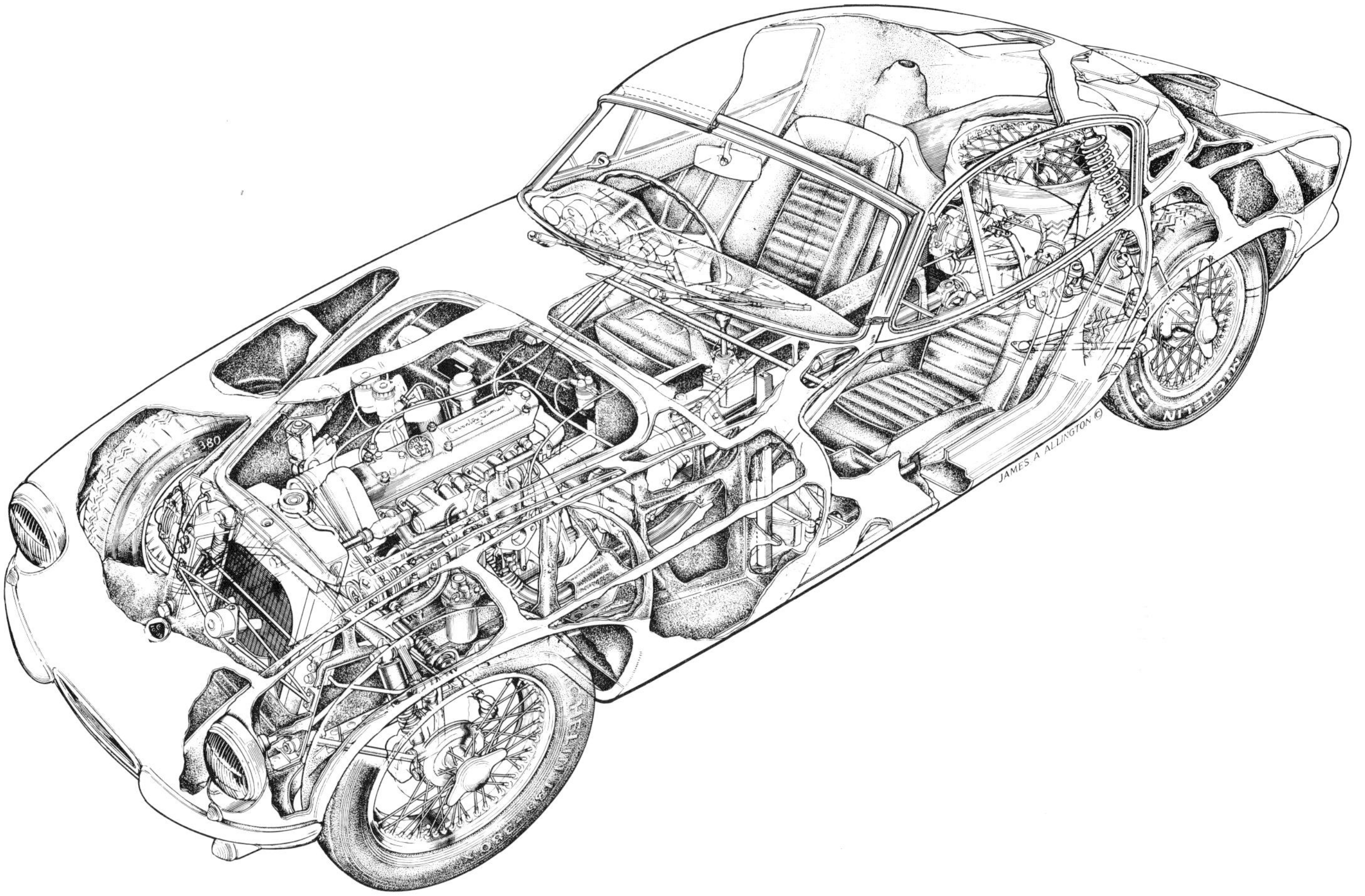

At its announcement, the Lotus Elite caused a sensation by its good looks, low build and exciting performance. It represented many 'firsts', not only for Lotus but for the industry. Shown here is the Series II, which differs most noticeably in the rear suspension. Notice the stiffening tube from the jacking point below the left door hinges, up the windscreen pillars and down the opposite side. The bulge in the top radiator hose is the thermostat housing, since the Climax engine did not have the normal integral housing provided.

accept the taller engine.

Development of the shape and the details of internal trim and appointments were being carried out on the drawing board and with the clay model. Meanwhile a 'mechanical prototype' had been made to test the suspension, steering and drive systems proposed. Since at this time there was no body unit available on which to assemble the mechanical parts it was decided to make a steel frame for test purposes. The layout which allowed the other components to be as near as possible to their anticipated form was a backbone frame of sheet metal construction. It was for these reasons that what has now become well known as the Elan chassis was conceived.

Only when the opportunity to incorporate the larger and much more powerful engines presented itself was the decision taken to make the backbone a permanent part of the structure. One of the supporting reasons was that the Elan was to be an open car. The Elite shell had had the benefit of the roof to complete its structure, but in the Elan the necessary strength had to be achieved without it. The team felt able to do this while the car was to have its original form, but knew that the engine change completely transformed the car's potential.

The mechanical prototype, meanwhile, was clocking up miles and discovering the 'bugs' in the initial ideas. Because the final body was still being designed, a body of some sort was needed for road

ELAN

testing. At that time a company called Falcon offered a range of glass fibre body shells for the special builder. These were fairly crudely finished but provided a sound basis for those building a sports car from components obtained elsewhere.

Because many specials were made from the early Ford 10 chassis the Falcon range included bodies designed for easy use with them. It was one of these that was chosen to clothe the Elan prototype.

The average builder, using many of the original Ford parts, made a car higher and longer than the Elan was to be. It was therefore necessary to adjust the shell to fit the Elan dimensions. Albert Adams, Lotus' leading glass fibre expert of the time, (and now a respected proponent of the exploitation of the modern composite materials that Lotus have been developing), took a horizontal slice out of the Falcon body to reduce its height by about six inches and shortened it just behind the door aperture by some ten inches. The body was re-bonded together with glass fibre and resin. It was not painted, since the glass fibre was self coloured in blue and was felt good enough for the purpose in hand.

On the road the car looked like a scruffily finished home-built special and lacked the comfort that even the most casual special builder would include. The floors around the feet of the occupants were made of plywood, bolted to extensions welded to the chassis. So as to make the body easily removable for access to the experimental chassis, the vertical ply panel behind the front wheel was not sealed against the wheelarch and left a gap of about two inches. This meant that if the road was wet a continual cascade of gritty water came over the top of the panel onto the feet of those inside. Since the floors were not sealed at the sides either, the water escaped readily enough, but those who drove the car regularly took to wearing rubber boots for the purpose.

Looking a little more closely at the Elan chassis, it can be seen that it has much in common with a monocoque, although it could never seriously be called one. Just as a monocoque depends for its strength on the thin shell of material from which its outer surface is made, the Elan chassis depends upon the sheet metal from which its outer surface is made. The difference comes in that a structure can only be thought of as a monocoque when the shell is at or near the outside of the object for which it provides the strength. The Elan chassis, of course, is largely surrounded by the rest of the car and so is more properly described as a backbone chassis.

The structure consists of a main central box beam of 'top hat' section closed in at the bottom to become a full box. This box form is very strong given the conditions that its cross section shape is maintained and that the loaded flat areas are sufficiently stable to avoid crumpling under load. This type of deformation is sometimes called 'oil canning', although in these days of oil being supplied in plastic bottles the term is perhaps less clear in meaning.

Distortion of the cross section of a box beam is ideally controlled by closing its ends with flat plates like the ends of any box. This cannot be done, though, if the ends have to be open to allow the passage of the drive shaft. The end plates, therefore, have openings which are reinforced by having flanged edges. The stability of the main surfaces depends on their width, so reinforcements may be added if need be.

At each end of the central beam member there is a sheet metal fabricated fork, the front fork being large enough to accept the engine and gearbox and that at the rear of a size to incorporate the differential unit. To stabilise the fork, its ends must be tied together. At the front this is done by another box member which, in true Lotus style, serves several purposes. It strengthens the arms of the fork member, provides mounting points for the front suspension lower wishbones and acts as a reservoir for the vacuum system for headlamp raising and lowering. A tower is welded to its outer ends which carries the upper wishbones and the spring/damper unit.

Lotus did not find quite so many jobs for the rear crossmember. It braces the rear fork, supports the differential unit and carries the towers for the Chapman strut top fixings.

A strong but shallow channel forms an extension to the underpart of the main box beam. This member carries the rear wishbone mounts and provides attachments for the two torque rods. To the untrained eye these rods seem to just stop the differential from swinging back and forth on its upper

ABOVE LEFT **The Lotus Elan. A contemporary press photo showing the Elan as first released. The engine was of 1500cc and the wheels bolt on. Later cars had the nominal 1600cc engine (1558cc) and centre lock wheels. The front end of the boot lid hinges are shaped to provide a rear location for the proposed hard top.**

LEFT **Sideview of that first Elan. These early cars had side windows sliding vertically, hand operated by a finger grip at the top edge.**

rubber mounts. This, of course, they do, but although the rear wishbone mounts transmit into the car the drive thrust from the rear wheels, the diff torque rods react *more* than the drive thrust of the car into the lower channel. More than twice the thrust driving the car is applied to the extension channel under the differential unit.

The central box beam, as has been said, is of top hat section. The flange (the brim of the hat) turns out horizontally to provide body fixings. Since the body was originally conceived to be strong enough without the chassis, it is capable of adding considerable strength. But this is only effective if the body and chassis are sufficiently well joined. For this reason there are several body to chassis fixing bolts. These are situated on the flange, on top of the box beam (also serving as a pick-up for the ply instrument panel, which is a true structural member, loaded in shear) and at or near the front and

The Elan chassis. This shot shows the chassis prepared for the car's first appearance at the London Motor Show, in 1962. As can be seen a great deal of trouble was taken over it. Note in particular the differential housing in aluminium and the four rubber drive joints. The brake discs are mounted to the three bolts which attach the joint to the outer drive shaft. The handbrake is rod operated through a mechanism called a whiffle tree, which shares the rod tension equally between the two brakes. The insulators at the tops of the struts are Lotus' first ever purpose-designed rubber mounts. They are in fact adapted from a proprietary range, but required a difference in stiffness between the fore-and-aft and the lateral directions. At the time, the cost of the special moulds quoted by the normal producers were beyond the Lotus resources, so they made them themselves.

ABOVE **Front view of the Elan show chassis showing engine installation. An engine mounted fan is shown although it rarely featured on production builds. The carburettor trumpets are at full length here, although for production they were shorter to fit inside the airbox. The access holes in the chassis have flanged edges for stiffening.**

rear suspension mounts.

Fixing a glass fibre moulding to a steel structure was a topic about which little was known in 1961-2 when the design was being developed. On the Elite, where suspension and engine mounts had been bolted to the shell, Lotus had used a steel insert known as a bobbin. These were made from hexagonal bar, with a wide groove turned in them to allow the glass moulding to grip them. Experience had shown that this form of bobbin had two main faults. Firstly they did not have enough resistance to turning within the moulding, so that if a bolt had rusted into the bobbin then the body would inevitably be damaged in attempting to remove the bolt. Secondly, and more important from the point of view of safety, they were too small in proportion to

BELOW **For comparison, the Excel chassis: nearly twenty years after the Elan, (here viewed from the front), the similarities are evident. The chassis is more robust, if only because the car is heavier and more powerful. By this date Lotus chassis were galvanised (zinc coated) for corrosion protection.**

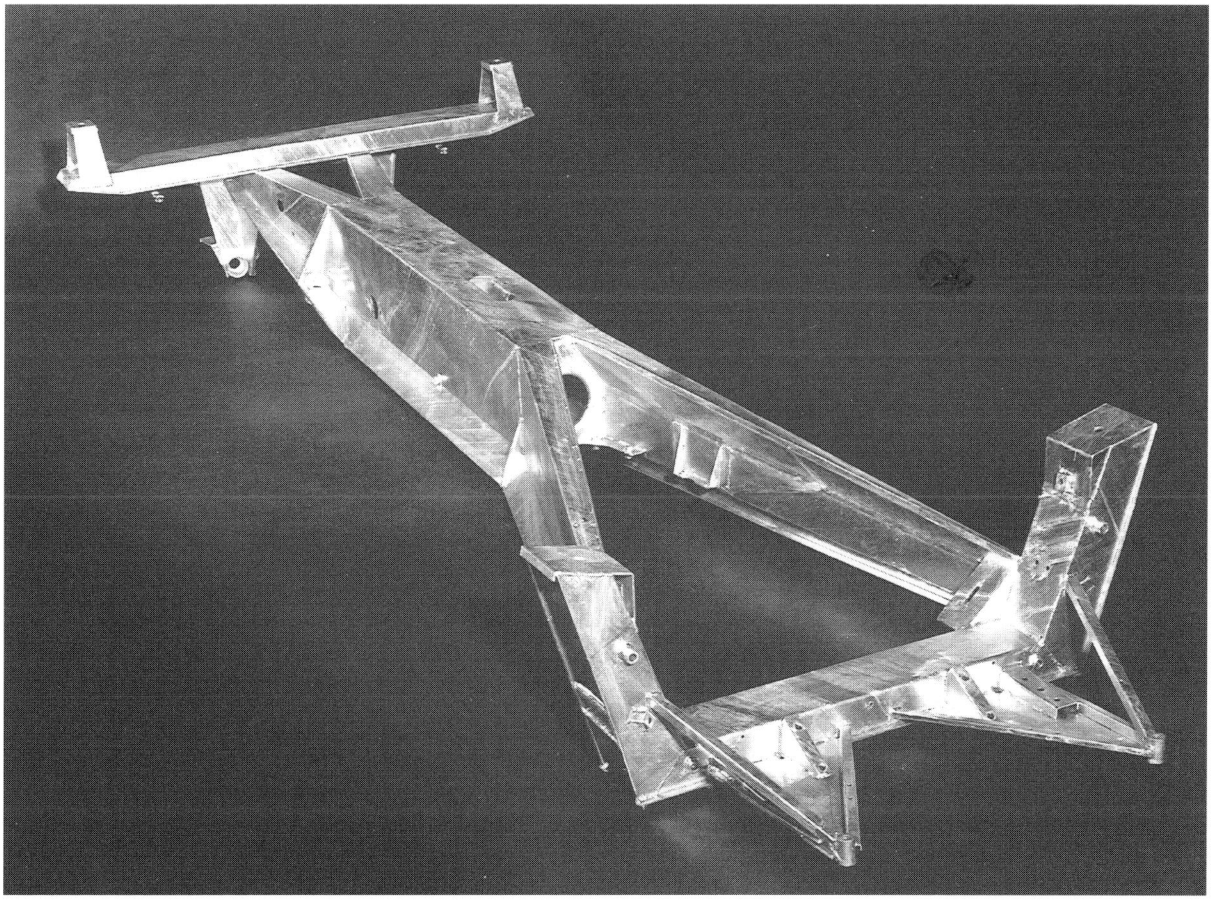

the loads they had to carry, and therefore were liable to tear out in operation.

Chapman's engineers were set the task of devising a better bobbin. Various styles and proportions were tried by moulding them into test glass fibre components. These samples were taken to the materials laboratory at Enfield Technical College for destructive tests. Out of this work came a design of bobbin cast in a slim oval shape, which gave very great resistance to turning, and having quite high flanges to spread the loads into the material as much as possible.

The new bobbins were to be cast in aluminium in order to reduce corrosion problems. Where fixings were to be used between chassis and body there was a particular problem in accommodating the position errors between the two parts owing to the

ABOVE **The DeLorean Car. The much discussed DeLorean project involved Lotus Cars as consultants from the early days. Quite apart from the financial and political complications, the project had its technical difficulties as well. If Chapman had had his way the car would have been very different, since he considered that many of the design features already pre-defined were quite wrong.**

RIGHT AND OVERLEAF **While employed by Lotus in the 1960s the author, with Albert Adams, carried out some tests to determine the best form for the bobbins which were bonded into glass fibre structures for the attachment points. This report shows the results of the work and even today's bobbins have essentially the same form.**

TEST REPORT NO. 6 10.4.61.

Prepared by H.H. Haskell.

REPORT ON TESTS CARRIED OUT ON LOTUS BOBBIN TYPE FIXINGS FOR RESIN BONDED GLASS FIBRE:LAMINATES:

Circulation:
Bristol Aeroplane Plastics:-
Mr. W.R.Farnes, O.B.E. - Director
Mr. J. Branston - Works Manager
Mr. J. Rees - Production Manager
Mr. P.Curnock - Foreman.
Mr. J. Britten - Chief Inspector
Mr. A. Baker - Chief Engineer

Mr. A.C.B.Chapman - Managing Director Lotus
Mr. P.C.Fish - Chief Inspector
Mr. T. Shann - Service Department
Mr. R.P.Hickman

OBJECT: To test the suitability of bonded-in metal bobbins as attachment points in Fibreglass structures.

BOBBINS: Two types were tested. Type "A" (Oval) made of aluminium

Type "B" (Hexagonal) made of steel

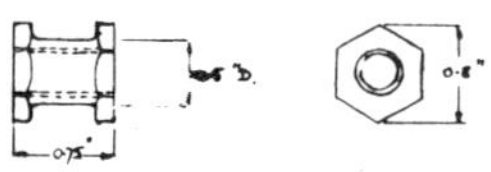

BONDING: The bobbins were bonded into the fibreglass test pieces in such a way that the glass strands of the laminate continued up the sides of the bobbin to form a sound fixing. (See fig.1) Various ways of achieving this end are discussed under "Conclusions."

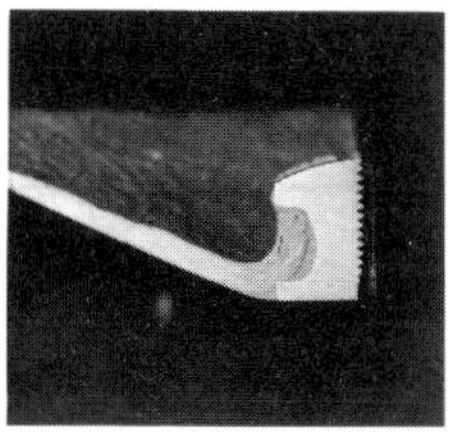

RESULTS:

TEST 1 - (Tensile) Oval bobbins:

Sample 1 (T.1)

Failing load 0.55 tons. At this load the bobbin was gradually being pushed out of the fibreglass. Failure was not sudden.

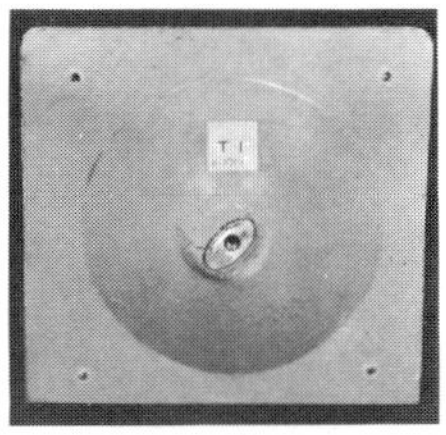

Sample 2 (T.2)

Failing load 0.44 tons. A failure exactly similar to T.1, but the bobbin was not pushed as far out of the fibreglass because the load was not applied for such a long time.

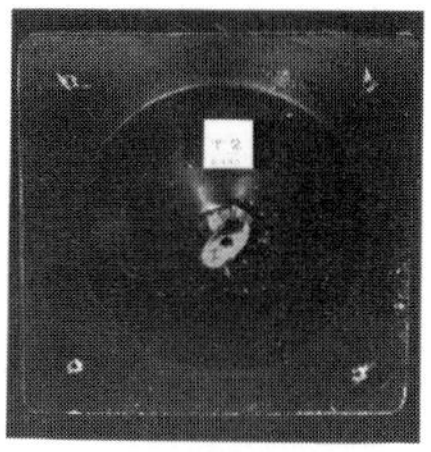

Sample 3 (T.3)

Failing load 0.565 tons. Similar to previous tests but again a shorter application of load.

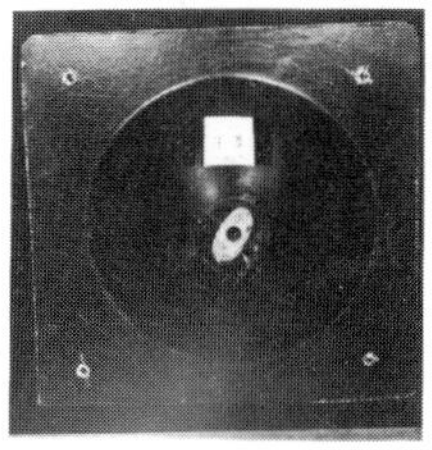

Page 2.

TESTS: Three types of test were carried out on samples of each variety of bobbin.

TEST 1 - Tensile Test:

A load was applied to the bobbin so as to pull it straight out of the laminate - this was actually done by mounting the test pieces face downwards over a hole in a wooden block and loading the bobbin from behind.

TEST 2 - Shear Test:

A load was applied to the bobbin so as to tear it sideways out of the fibreglass - this was done by mounting the test piece in the apparatus shown in Fig. 2. The test piece (c) is fixed to the frame (b) by ¼" screws through the 8 holes in the flanges of the frame. Bar A is then screwed to the bobbin with a suitable Bolt so that it lies on the opposite side of the frame to the bar which is part of the frame. The two bars are then fitted into the jaws of a tensile testing machine.

Fig. 2.

TEST 3 - Torsional Test:

For this test a stud was screwed into the bobbin and a nut run down the stud and tightened against the bobbin - the nut was then turned with a torque wrench until failure occurred.

TEST 1 - (Tensile) Hexagonal bobbins:

Sample 1 (T.4)

Failing load 0.4[illegible] tons. A gradual pushing out of the bobbin.

Sample 2 (T.5)

Failing load 0.[illegible]1 tons

A similar failure to T.4 but carried to a greater extreme.

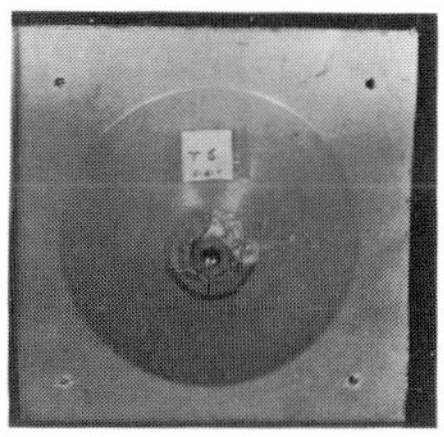

Sample 3 (T.6)

Failing load 0.58 tons Exactly similar to T.5 but carried still further. In this test and the previous one the missing pieces of fibreglass were removed to see the internal damage.

Sample 4 (S.4)

At a load of 1.25 tons the 5/16" UNC H.T.bolt screwed into the bobbin sheared.

TEST 2 (SHEAR) Hexagonal bobbin:

Sample 1 (S.5)

Loaded to 1.25 tons at which the test piece tore away from the frame.

TEST 3 (TORSIONAL) RESULTS:

Oval Bobbins.

Sample 1 (To.3)

This is the same sample as for Sheer Test 1. (S.1) At a torque of 48 lb. ft. the thread in the bobbin stripped; no movement of the bobbin in the fibreglass was apperent.

(See Page 5.)

"CONCLUSIONS".

From the figures and results obtained, it can be seen that the strength of this type of attachment is as great as could ever be required, since it is stronger than both the laminate and the bolt screwed into the bobbin.

The strength, however, will only be adequate if the bobbin is properly bonded in to the fibreglass. The method used for the test pieces was as follows:-

> The bobbin was bolted into the mould before application of the gel-coat. The gel-coat was then applied, care being taken not to allow any resin to get on to the sides of the bobbin. Four layers of 1 oz. chopped strand mat were applied next, each one separately. Cross cust were made in these layers to fit over the bobbin as shown in Fig. A, the cuts in each layer being arranged to be at 45° to those in adjoining layers. As each layer was applied the four points resulting from the cross cuts were worked into the groove in the bobbin to hold it firmly in. There were no glass filaments wrapped around the bobbin.

Alternative methods of laying-up satisfactorily are shown in the accompanying drawings.

(For Fig. A see page 9.)

TEST 3 (TORSIONAL) Hexagonal Bobbins:

Sample 1 (To. 1)

This bobbin started to turn in the fibreglass at a torque of 44 lb. ft.

Sample 2 (To 2)

At a torque of 48 lb ft the ⅜" UNF H.T.Bolt sheared, the bobbin having moved only very slightly.

INSTRUCTIONS [illegible] BOBBINS.

RECOMMENDED BOBBIN LAY[illegible]

Note cross slits cut alternately square and diagonally.

Fig. A shows system used for test samples.

NOTE: If necessary continue to add (say) 3-4" squares until laminate around bobbin reaches tightly under top shoulder to prevent bobbin pulling out. (See Fig. B).

Fig. A

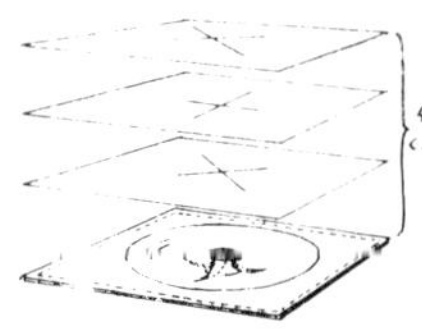

Fig. B

ADDITIONAL LAYUP TYPE 1: (All types of bobbins):

In addition to the above recommended system it may be advisable to apply 3 or 4 windings of tape, woven rovings, or a spiral of ordinary rovings around the bobbin AFTER completing the basic layup, so as to increase the "Bursting" strength around the bobbin. (This applies particularly where high torque loads are expected as when tightening a stubborn bolt when resin has got into the threads).

ADDITIONAL LAYUP TYPE 2: (Hexagonal bobbins):

Where the bobbin is too long to allow the pointed flaps on the basic layup patterns to reach under the top shoulder, the following additional pieces should be used, these to be inserted between alternate "layers" of the basic layup, as shown in Fig. C.

Fig. C

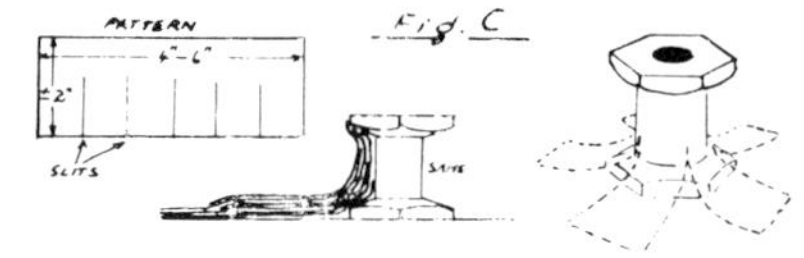

tolerances of manufacture. Every manufacturing process, however precise it may be, has some amount of variability in the parts produced. If two parts are to be assembled together at two points, the distance between the holes on one of the parts must match the distance on the mating part, even when one part comes out at maximum size while the other comes out at the minimum. The manufacture of a largish light sheet metal fabrication is not normally a very precise process. Moulding of glass fibre shell gives rise to considerable variation too, and when the fixing points are not two but multiple, the problem is increased proportionately.

To allow for these tolerances, Lotus devised an extra large bobbin based on the smaller one, with a large clearance hole to ensure that the bolts would line up whatever the tolerance build-up between body and chassis might amount to. At the same time as determining the best form for the bobbins, the Lotus engineers were able to research and define the best way to 'lay up' the moulding in their vicinity. The work done at this time (early 1961) formed a valuable foundation for many designs subsequent to the Elan and the principles then laid down were still being used in the 1980s.

Here and overleaf: progress toward monocoque. The 21 was really the bridge from the 18 to the 1.5 litre formula. It had the four cylinder Climax engine from earlier years, since the V8 was not yet available. It had for the first time inboard front suspension units but used uprights adapted from the faithful Triumph upright used since the days of the Eleven, and the rack was still in the low position. Notice the immense fuel tank along the left hand side and crossing under the seat. This tank was typical of the time. It was a nightmare to make and worse to make leakproof, shaped as it was around each chassis tube. A secondary tank behind the right hand front wheel (connected by a tube to the main one) increased the total capacity.

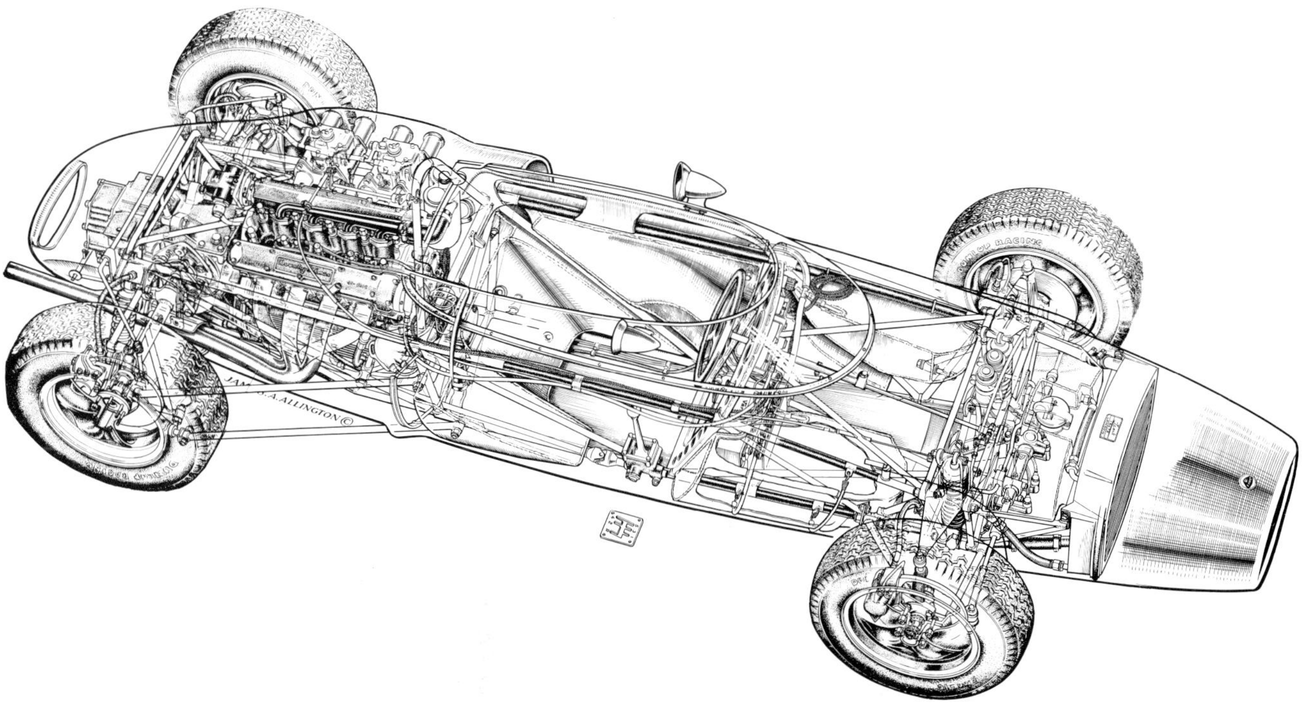

JAMES.A. ALLINGTON ©

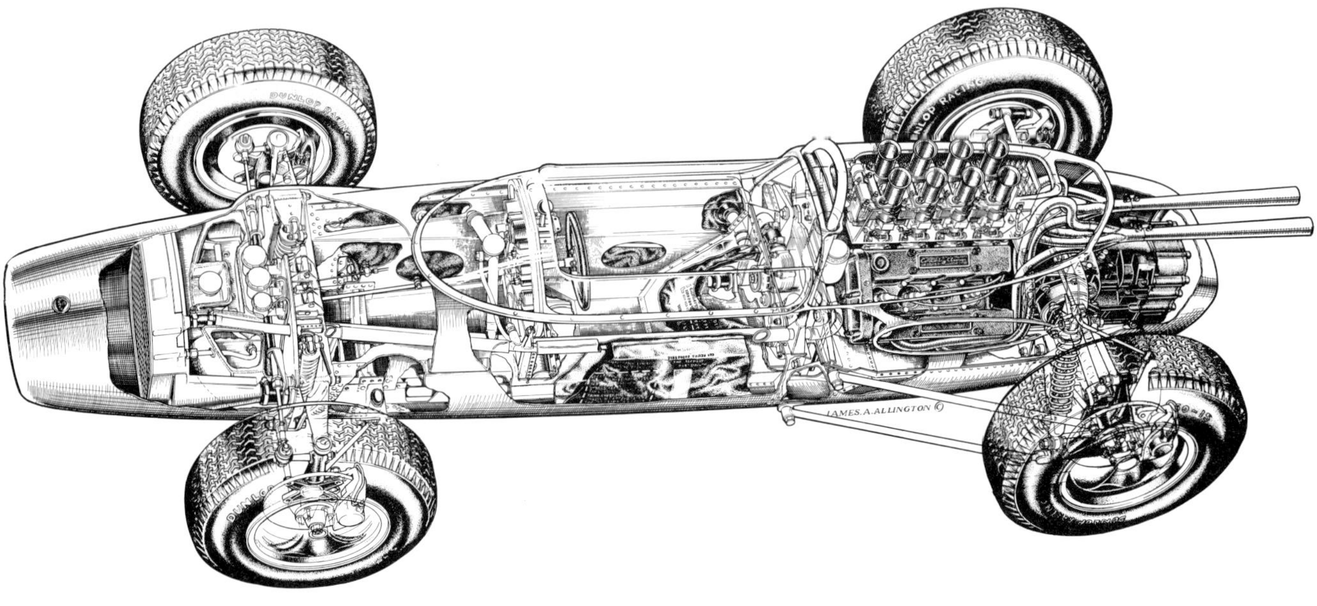
JAMES.A.ALLINGTON ©

ABOVE LEFT **It can be seen in this cutaway drawing by Jimmy Allington how the 24 links the 21 and the 25. The chassis is tubular and the rack is now raised to the plane of the upper suspension arm. Notice also the gearshift low on the driver's left. The pedals are bottom-hinged. These points are the main differences from the later 25. All the suspension components are virtually interchangeable, although the tyres are one size smaller in cross section.**

LEFT **Looking at the 25 with its monocoque shell it can be seen how Chapman, while creating a much better structure, was able to take advantage of it in a greater elegance and cleanliness in other areas. The gearshift has now gone to high on the driver's right; the pedals are now top-hinged (pendant) as this allowed neat integration into the front suspension sub-frame. The fuel is now held in bag tanks in the sides of the structure, with an extra bag behind the seat; filling is via a single filler ahead of the instrument panel. The radiator is changed to allow for a neat air duct for cooling the suspension units.**

ABOVE **The 33 was the ultimate in the 1.5 litre formula series that started with the 21. It was also the ultimate development of the revolutionary 25. Compared with this car it was slimmer, lighter and simpler. At last the wobbly web wheels have been replaced, in adopting the newer, much wider, tyre profiles. It does not differ from the 25 in any radical way; it is a 25 'only more so'.**

8 The Lotus Cortina

DUE TO FORD'S sensitivity over the association of its name with that of other manufacturers, the car the world has come to know as the Lotus Cortina (one of the classics of the 1960s) was christened the 'Ford Consul Cortina Developed by Lotus'.

Through his innovative manufacturing methods Henry Ford achieved his aim of popularising the motor car, but this was gained at a serious cost to the image of his company's products. Certainly Ford held little appeal for the motor sporting fraternity.

As part of Ford's campaign to improve its image and attract a part of the car-buying public that it had previously neglected, Ford decided to enter motor sport. One way the company did this was to sponsor rally competitors and later to run teams. In the early 1960s production saloon car racing was a very popular part of the sport and attracted teams from many manufacturers.

At this stage Ford had not gone beyond sponsorship of private entrants driving their own cars, but, by 1962 was ready to enter in a more serious way.

After the company's usual extensive research, the particular sector identified was the class for series production saloons, the international 'Appendix C'. Saloon cars were raced in three classes – standard, modified and no holds barred.

The first class allowed virtually no modifications to the vehicles and consequently attracted very little interest. In the second class, limited modifications were permitted so that the cars seen racing really were quite close to those that the spectators could buy. The final class allowed (almost) any modification and was contested by such animals as Jaguar-engined Minis – not an appropriate class for serious manufacturers to promote their products. Ford opted for the middle classification of racing.

It was then a matter of considering the car model range and the engine size which best sparked the public interest. To go into the small car class was out of the question since Ford had no Mini-sized car. In the three litre class Ford could have entered with the Zephyr range, but this range was not attracting much international interest and in any case these cars were not normally significantly faster than the best of the big-engined Anglias, and were unlikely to be able to match the Jaguars.

Having thus settled on the modified 1½ to 2 litre class, Ford then had to choose its manner of attack. As we now know, the company chose to collaborate with Lotus. The Lotus Cortina was conceived roughly as a Cortina made to run in the 'no holds barred' class, but manufactured in sufficient quantity as to qualify as a production car in its own right.

When Chapman had determined the specification of the 'ultimate' racing Cortina it then had to be de-modified by the right amount to ensure that in its fully modified form it would come in the so-called ameliorated class.

Production Lotus Cortinas were to be pitched at such a level that with the permitted modifications of Appendix C of the International Sporting Code they became 'Cortinas with unlimited modifications'. In other words, Production Lotus Cortina equals freely modified Production Cortina.

A company like Ford is very methodical in its approach to any new product. Each product on which work is done (even if it never reaches production) is defined at its outset by a series of Product Letters giving its market objectives. These continue through the development of the product and beyond, refining the specification as it matures.

Ford released this picture of the prototype Lotus Cortina under test at Silverstone in 1962.

The Lotus Cortina started life as boldly as one might wish. Some wag at Ford established the series of Product Letters for the Lotus Cortina by publishing the first one under the reference HM 1 (HM – hairy motors). From this beginning was born the car which has since made its place in motor sport history.

Initial negotiations were carried out confidentially, but once the project was started, meetings were held between the two companies involving widening circles of people. There were also a series of visits to each other's factories and engineering departments. Some doubt was cast in the minds of the Ford engineers, used to seeing several hundred engineers and draughtsmen in their engineering centre, to see the Lotus drawing office with a total of eight! At the time Lotus had only a small development shop which served for almost all prototype manufacture. This shop was well equipped for making racing cars but not for handling a passenger saloon. There was no lift for raising the car for work on the underside. Racing cars were light enough to be lifted by two or three people onto trestles when needed, or at most using an engine crane.

The method of working was that Chapman, who had had drawings of the car some time before, produced schemes outlining the suspension geometry, driver seating positions etc., as he envisaged them. These drawings, which Chapman produced precisely to scale and incredibly neatly, contained all the keys to the Chapman ideas to be incorporated in the design.

It was this type of drawing that he always used to pass his guidelines to his design team. The drawings would be accompanied by notes in his fine hand writing. In the case of the Lotus Cortina many of these notes had already been passed to Ford before the project started. They included the requirement for aluminium body and gearbox parts and diff housing. Without Ford's undertaking to provide these, Chapman would have been most reluctant to undertake the task. Remember that he was in effect promising to produce a world

beater. To do this he demanded that it should be just as he specified.

It must be remembered that this product was not a Chapman design and that most of the detail work was undertaken by others. However, the Chapman vision was very much in evidence. He manipulated the early negotiations with Ford in such a way that he could specify the car to fulfil the original requirements while at the same time so surpassing the minimum necessary that the product became a classic.

The most noticeable of these changes were the use of aluminium for the bonnet, boot and door pressings and for the clutch housing, gearbox extension and diff castings. Although, when Chapman originally asked for these changes the Ford senior management had agreed them, when the engineers from the Aveley Engineering Centre came to carry them out they quite naturally questioned the costs involved. To do things 'correctly' the tools made for forming these parts should have been remade to suit the different material.

In a press tool, in which body panels are made, the shape of the tool differs from the shape of the finished panel to allow for the spring-back of the material after forming. Sheet aluminium has a different springiness from sheet steel and theoretically demands new press tools. Similarly, since casting tools differ in form from the parts they make to allow for the shrinkage of the material (aluminium and iron have different shrinkage rates), new casting tools were required. Further to this, the finishing processes were different for aluminium panels and would need to be new. Also machining settings for aluminium are different from those for iron and so would require different arrangements.

Once the Lotus Cortina appeared on the race tracks its reputation for results – achieved in spectacular ways – was soon established. Jim Clark was famous for his three wheeled driving, but was by no means alone. Once, when a remark was made to him how extreme this seemed, he replied: 'You should be *inside*'.

All these difficulties are real, not invented, as Chapman well knew. If new tooling was not made, the parts concerned would come out different from the heavier parts they replaced. The Ford engineers, thoroughly professional and trained to do everything by the book, took the view that they were required to make the lightweight parts to exactly the same dimensions as their original counterparts. Chapman argued, however, that it did not really matter if the parts were away from the original drawing provided they fulfilled their functions. This was a case of not being bounded by known limitations, an example of the experts 'knowing what could not be done'.

Even while accepting Chapman's reasoning, the Ford people were not convinced that the normal tooling would work acceptably with the lighter and softer material. Chapman had to insist on actual trials being done.

One part of Chapman's specification was the vertical setting of the front wheels. This was to be achieved by lengthening the lower suspension link (in Ford terms the track control arm). The early drawings had shown this as a fabrication from sheet metal with blocks welded in for ball joint attachment etc. This construction was also used on the car provided for conversion as a prototype. A pair were duly modified by Lotus by cutting and re-welding – a process which they themselves regarded as highly unsatisfactory.

The Aveley engineers were quite horrified that anything so nasty should be done to a safety critical component! Fortunately, however they were able to help, since they knew of a more recently designed forged version intended for heavy duty applications. This track control arm was very welcome to the Lotus team since they were already quite at home with reshaping forged parts. They had been using Standard/Triumph parts with re-formed steering arms ever since the Eleven. A jig was therefore made on which the forged track control arm could be fixed while the neck between the anti-roll bar fixing and the outer ball joint boss was heated and straightened slightly (even though the experts said you could not do it).

Chapman had laid down 5½ inch wheel rims for the new car. This width, on the 13 inch diameter of the Cortina wheel, was quite unknown to Ford, although Jaguar used it in 15 inch diameter. The Jaguar rim was of a thickness appropriate to the weight of their products but Chapman had specified much lighter 14 gauge (0.08 in). For the prototype, Dunlop, who made the Jaguar wheels, made up rims by

Chapman was well aware of the Lotus Cortina's cornering style, but took the view that there was no point in making an effort to keep the inner wheel on the ground. It was important to keep both rear wheels down, though, as they provided the traction. For this reason the roll stiffness at the front was made much higher than that at the rear, with the result seen.

cutting down some of these to the smaller size.

In subsequent negotiations on the production rims, George Adams, then Dunlop's chief designer and well known throughout the industry, simply refused to make rims that thin. A compromise was reached on 13 gauge. (A few years later, when the Mk 2 Lotus Cortina was to be made by Ford and they were to make the 1600E model using the same suspension, Ford remade the wheel with an even thicker rim again since the 13 gauge was still not robust enough by their standards). The first prototype car was used for a demonstration to the Ford management of the car's potential, as a confirmation that the original promises were being fulfilled. This car started life as a 1200 cc two door Cortina. It had, of course, an all steel body construction, but for the time being this was not important. In its reincarnation as a Lotus Cortina it was fitted out with two racing bucket seats, as used on the Lotus 18, mounted by a tubular steel frame to the normal seat runners, and trimmed to match the car interior in iridescent blue.

A demonstration day was arranged at Silverstone which was rented exclusively for the purpose. The test day was a Monday, and as the car had been finished late on the Friday night, Lotus had two whole days to get on some miles.

One of the known difficulties was that the Ford rear axle, designed years before for a much lower powered engine, simply could not stand the torque of the twin cam engine unless run in. It had been found in tests with the Anglia which had been used as a test bench, that a well run-in axle coped satisfactorily, but that one had done some 30,000 miles with an ordinary engine first. Ford were confident that production axles would be protected from new by a special initial fill oil (not then available) and this proved to be the case.

Several Ford directors and other senior managers were present for the demonstration and Jim Clark was there to drive. Chapman was there as was Mike Costin, then still part of the Lotus team. Of the Ford men present some were very much motor racing people, some even raced themselves. On the other hand some had no more contact with the sport than driving round the North Circular Road.

Jim Clark had got in a few practice laps beforehand and, this being only about the second outing of this first Lotus Cortina, Chapman was most anxious to see how his theories were working out. Chapman also tried the car, turning in laps scarcely slower than Clark and still better than the lap record of the time. Chapman felt that the car might be improved with a little more rear-end weight. He therefore tried a few laps sitting on the rear seat (with no seat belts) while Clark drove. When they stopped those present crowded round to get his impressions, which were that 'He stays calm and keeps his foot down when I would have lifted off'.

When the time came for the demonstration everyone stood around while Clark did his stuff and showed that Ford's new toy could deliver better performances than previously possible for the racing class aimed at. At some point Chapman suggested that they might like to experience a few laps with Clark. One or two of the Ford people volunteered and all of them looked somewhat shaken when they finished. One in particular, incidently not a racing buff, remarked that he first wondered when Clark was going to stop accelerating – and then realised he was not.

Towards the end of the day, when all the important work had been done, Clark was just completing a lap when the nearside rear wheel started to come away. On these axles the half shafts are held in only by the fact that the bearing is press fitted to it, with a backing ring. What had happened was that the shaft, subject that day to cornering forces that it was not developed for, had started to pull out of the bearing. Clark came to a halt alongside the spectators with the wheel about a foot out from the car. The car was jacked up and the shaft removed to be shown to George Baggs, head of Special Vehicle Engineering, who was responsible for the technical side on Ford's behalf. He subsequently arranged for heavier back-up rings to be made for later builds. The offending bearing was forced back into position and Chapman drove the car back to Cheshunt, more worried by the noise from the final drive gears which were by then loudly proclaiming the abuse they had suffered.

One of Chapman's stipulations was that the engine should be fed with cold air, rather than sucking in under-bonnet air. It is well known that an engine produces more power using cold air than warm and this is because the colder air has a higher density. This means that as the engine pulls in a fixed maximum volume of air it will get a larger quantity in this volume if the density is higher. The air brings with it a due quantity of fuel to produce the maximum power.

Ford's practice (as most other makers') was to simply let the engine draw on under-bonnet air. This has the benefits of simplicity of installation and also means the engine will be more economical on the less dense warm air. Since the twin cam engine was fed by two twin choke Webers it was not entirely evident how to provide a cold air supply.

On sports and racing cars it was usually easy to provide an opening in the bonnet to allow the air to flow in close ahead of the carburettors, but with this car that was not possible. In addition, open carburettors make a great deal of noise, so full intake silencing was needed. The solution was to provide a closed housing for the air intakes fed by a flexible pipe to the space ahead of the radiator. An air cleaner/silencer similar to the normal one was mounted in this area on its side.

Suspension problems

Much has been written in the specialist press about the reinforcing tubes in the boot space of the earlier Lotus Cortinas. The reason they came about is that Chapman did not realise the effect of the loads when he specified the rear suspension system. At this time he had no experience of the design and construction of this type of car body.

The chassisless construction of the modern car was of great interest to him as a lightweight structure and he had studied it at a distance. During the development of other products, notably the Elan, such structures were sometimes used as comparators for Lotus' own performance at creating light and stiff frames.

The Cortina shell was particularly light for that time and its overall figures of torsional stiffness were good. But knowing what might be called global facts such as these did not ensure that Chapman knew the details of how it was achieved. To make a structure light it is necessary to ensure that it is no stronger (i.e. heavier) than necessary.

The structure that Ford had developed was not light by accident, but as a result of careful study of the loads involved and the way the structure resisted them.

Semi-elliptic or 'cart spring' rear suspension systems support the vehicle at four points, the front and rear attachments of each spring. At each point there is an upward force on the body of about a quarter of the weight of the rear end of the car. What is more, the front of the spring pushes

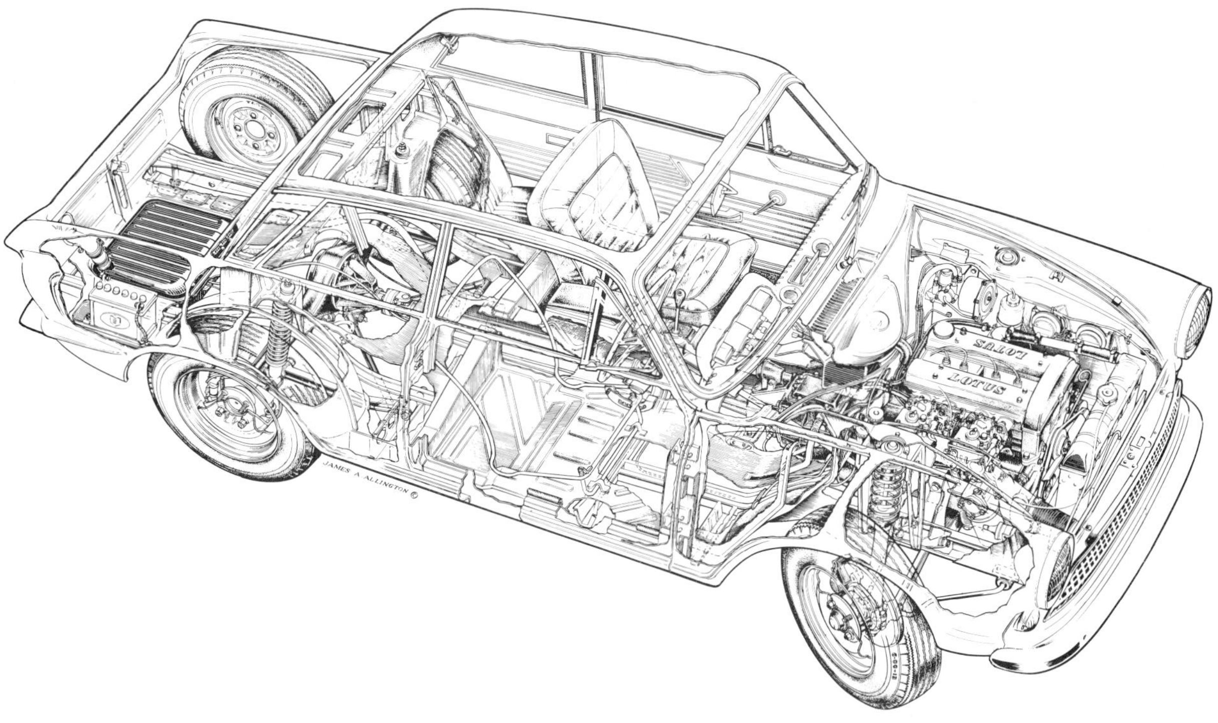

Jimmy Allington's drawing of the Lotus Cortina is of one of the production prototypes. Note the absence of rear end internal bracing. Also the A frame centre pivot is of the early threaded type. There is a spacer in the driveshaft also, which indicates a provisional design.

upward ahead of the axle, under the rear seats, while the rear of the spring pushes upward under the boot. Both of these places are exactly where support is needed since they are loaded areas. Ford's body structure was designed taking the suspension system into account and therefore there was very little need for great beam strength over the rear axle. Indeed it is quite difficult to provide bending strength at this point since the structure is shallow – the underside rises over the axle and the upper profile comes down with the rear windscreen.

Removing the four supporting points and replacing them with two, thrusting upward twice as hard, and at the weak point of the body away from the loaded areas, was guaranteed to create the most unfavourable situation possible. This is clear with the benefit of hindsight. But at the time, with minds preoccupied more with problems of handling on the one hand and of the logistics of the two firms working together on the other, this fact was missed. It was missed not only by Chapman, but also by his team and by the people at Ford, who were very experienced on this type of product.

The points where the loads were applied to the body were, it is true, those strong points provided for shock absorber attachment. Shock absorber loads, however, are very different from suspension loads for two main reasons. Firstly, they are intermittent (on a smooth surface no force is transmitted at all) whereas suspension loads are present at all times. Secondly, shock absorber loads are generally lower in the upward direction than in the downward, due to the bias built into the dampers themselves, if anything, therefore, easing the bending loads induced in the body.

The net result of this was of course that after a few road miles the tail of the body would start to droop under the combined effect of the concentrated upward load and the extra boot load of the large spare wheel and the battery. This droop showed itself on the outside of the car by the appearance of a ripple in the panel above the wheel arch.

A possibly apocryphal story has it that when Chapman first saw these ripples he assumed someone had been using the car without permission and had damaged it. The Ford people though, recognized its true significance as a real problem and not just an isolated case of damage.

Ford were very quick to come up with a drawing of the fix they could incorporate in the body during manufacture at Dagenham. To Lotus' surprise, Ford proposed the tubular bracing which became a feature of the boot space until the suspension was revised after a year or so of production.

Chapman was concerned at the outset of the project that because of its size and, more particularly, weight, the suspension frequencies were way above those he was used to dealing with. Suspension designers often use the frequency of the suspension as a means of comparing one with another. This is the bounce frequency of the vehicle on its springs – a softly sprung car has a low frequency and a high frequency gives a hard suspension (such as on a racing car of the last twenty years or so). At this time Chapman's success in racing had been achieved with relatively soft (low frequency) springing. The softness that is possible is dependent upon the combination of the suspension movement available and the weight (more correctly, the mass) of the vehicle.

Passenger cars, and the Cortina was typical, had a weight of perhaps three to four times that of the racing cars of the time, but the suspension movement is only about one and a half times greater. Thus the frequencies are inevitably higher by a factor of between two and three times.

It was clearly out of the question to ask for an independent suspension at that time due to its cost and also to the fact that Ford would not have wanted to deviate so much from the initial design of their car for what was in effect publicity for it. So Chapman looked to his most successful previous live axle rear suspension system – that of the Lotus Seven. The difficulties of translating the radius arm and 'A' frame from a very light two seater to suit a full four seat saloon were considerable. The tubular frame of the Seven had been designed around the suspension and consequently had strongpoints where needed; the axle beam was very light and everything was therefore easy to make work reliably.

Brackets to weld to the underbody reinforcements to anchor the 'A' frame were quite simple to conceive. The radius arms or upper suspension links could quite simply be designed to make use of the normal spring fixing points. If the attachments to the body were the easy part, things were harder in the design of the 'A' frame itself.

Tube was used for its well known properties, and a size was chosen appropriate to the loads anticipated. The end fittings had to be made from robust forgings. Since the roll angles of the car were to be much greater than for the Seven, rubber joints were not considered suitable. For this reason a pivot was designed into the frame.

For many years, as has been said, the front suspension upright from the Standard/Triumph small car ranges had been used on Lotuses. This upright had at its lower end a screw thread which served to carry the weight of the car and, turning within its threaded bronze lower trunnion, allowed for steering movement. On the Seven Series I and the Eleven there was a similar arrangement for the lower wishbone outer pivot, where the pin was threaded into steel bushes. This detail came from the Standard Eight and Ten of the 1950s. By the 1960s it had been replaced by a plain steel pin in plastic bearings. It was from this experience that Lotus drew in proposing the original threaded 'A' frame centre pivot.

The concept, while quite sound in principle, had two drawbacks. Firstly, to work reliably over a lengthy period it would have had to be made with great precision in hard materials. Secondly, being a metal to metal bearing it transmitted noise and shock excessively. The design that replaced it was produced by Ford after these snags had been shown up in testing, but too late for incorporation at the outset of production. Ford's new scheme tackled both of the above difficulties by using soft plastic bearings able to work with normal production tolerances.

An interesting insight into Chapman's thinking was, perhaps, that which involved the rear radius arms. On the Seven the radius arms were tubular and the rubber bushings allowed for the roll of the body. With the Lotus Cortina, as for the 'A' frame, more movement was demanded of the radius arms than could be afforded by the rubber bushes. The Lotus Cortina radius arms had initially been drawn – following Chapman's sketches – as made of tube. It was realised that they would be called upon to twist with body roll, which tube is not good at. The handling was therefore impaired by their torsional stiffness, which made them act like an anti-roll bar, causing unwanted weight transfer from the inner wheel on a corner to the outer. This had been noticed, remember, on the first test day. To allow an immediate test to be made before new radius arms were designed, Chapman suggested slitting the tubes lengthwise.

Although this destroyed the torsional stiffness of the tubes as required for a test, it would have resulted in a drastically reduced service life had it been used in production. A circuit test showed that the traction of the car was improved as anticipated, which proved that the original problem had been correctly diagnosed and the design of some new radius arms in channel section steel was put in hand.

The Lotus Cortina went on through one or two developments, including a divided drive shaft which greatly improved its smoothness of operation. When Ford introduced the MkII Cortina, however, they decided to terminate the sub-contract assembly carried out by Lotus and to bring the car to Dagenham, alongside the other variants.

All in all, this project considerably advanced Lotus' reputation in the eyes of a wider public and also gave the company valuable experience (and demonstrable success) with a different type of product. The later expansion into consulting could be said in some small way to be linked back to this time. Lotus also enjoyed a fairly stable income for quite a while without having to make much sales effort, since the company had the Ford dealer network working for it.

9 *Aerodynamics – or what to do about air*

CHAPMAN'S INFLUENCE, OR Chapman-inspired influence, as far as aerodynamics is concerned is extensive and varied. He was, as he was in any field, a follower of and improver on the ideas of others, as well as the creator of new ones. He turned his attention to such major matters as the overall airflow round a car and the development of ground effect vehicles as well as to relatively minor points such as radiator ducting. As with all things, Chapman's attitude was to pay attention to detail. He never let go a point of design until it was as perfect as he was able to make it.

With his earlier cars, because their speed capability was low, the importance of drag reduction of the body was also low. Chapman, with his aeronautical background and engineer's training, understood very clearly the effects of airflow on his creations: for any car, the total resistance that has to be overcome by the engine consists of two main parts – the air resistance and the rolling resistance.

The air resistance (the force with which the air resists the car's passage) increases with the square of the speed. This means that if the speed at which a vehicle is moving is doubled, its air resistance is quadrupled. The rolling resistance is in simple terms the friction in all the wheel bearings, the gearbox, drive and tyres etc. This force remains roughly constant with increasing speed.

The power required to overcome a resisting force is proportional to the speed of travel. Thus, if the force itself increases with the square of the speed, the power required to overcome it depends on the cube of the speed.

The table gives some values for a typical vehicle. Notice that at low speeds the power needed to overcome friction is the major force, whereas at higher speeds it is the air resistance that dominates.

Cars produced by Chapman from about the Mark III onward (for racing as opposed to trials) all showed attention to body drag reduction. This became evident to the untrained eye from the Mk VIII onward, where speeds in excess of 130mph were catered for. The Mks VIII, IX and X were all basically the same body and their aerodynamic form was developed by Frank Costin in his first foray into car body design.

Here, Chapman was not the major contributor, but he was the driving force behind the project and

Table illustrating how resistance varies with speed:

Speed	*Power absorbed by rolling resistance*	*Power absorbed by air resistance*	*Total Power required*
mph	**HP**	**HP**	**HP**
10	1	0	1
20	3	1	4
30	4	2	6
40	5	5	10
50	7	10	17
60	8	17	25
70	9	27	37
80	10	41	51
90	12	58	70
100	13	80	93
110	14	106	121
120	16	138	154
130	17	176	193
140	18	220	238
150	20	270	290

The first streamlined Lotus was the Mk VIII. Chapman raced regularly in this car, SAR 6. The engine was a 1500 cc MG.

had seen the importance of *outstanding* aerodynamics – as opposed to just good. When the car first appeared in public it was so remote from the conventional idea of a streamlined small sports car as to be quite a sensation.

This was perhaps the first sign of innovation in streamlining we were to see. At that time, one can perhaps now say that many designers were going for 'token' streamlining – with Chapman it was definitely the real thing.

Headlamps were provided (to make the car legal), but they were retractable. The car was not so sophisticated that their raising and lowering was automatic – far from it, the driver had to stop, raise the bonnet, pull up the lamps by hand and close the bonnet before they could be used.

However, the lights were there and were usable, but contributed no drag when down. What is more they were actually legal when raised, since the law required headlamps to be at least 22 inches above the road (to their bottom edge). Many subsequent sports racers ignored this.

Aerodynamic considerations on many cars were confined to providing smooth airflow over the body shape as dictated by the principal organs of the car. The shape would be made close-fitting to the necessary main parts and then shaped in the interests of minimising the drag. These cars, benefitting from Frank Costin's skills in aircraft aerodynamics and his understanding of how to use air to provide stability at speed, were equipped with fins behind the driver. Their purpose was to provide directional stability by ensuring that the centre of the side surface area was sufficiently rearward, rather like an arrow.

Streamlining and other air-flow related considerations were in evidence in many other parts of the design, including the panel that separated the under-bonnet area from the wheel arches, which was sealed with foam rubber to the underside of the bonnet as it closed. The front air intake was wide, cooling the radiator and, at each side, the front brakes. The radiator was fully ducted; from the opening at the front an aluminium duct expanded to mate with the outer shape of the radiator core, and behind, another duct carried the air away downwards to exit underneath.

The cars had a full undertray – in common with later models – but not in common with all their competitors. It was well understood that this ensured there was reduced air pressure under the moving car, so that to take in air from a region of high pressure at the front and exhausting it underneath ensured the necessary flow with the minimum of drag. The small size of the intake and the expansion of the inlet duct were calculated, taking into account the slower air flow through the radiator due to the resistance of the small passageways. As the air then passes through the core it is heated and expands, leaving the core at a higher speed. It is then accelerated further by passing through a narrowing outlet passage with its exit under the car. Thus, by careful design of the ducting, some 'jet propulsion' effect is aimed at. This is not to say that the car could be shown to be pushed along by the thrust so generated, but that such an approach must result at least in a minimising of the drag developed in the cooling system.

Although the cars were two seaters they had metal panels (tonneau covers) over the passenger seats, together with a single windscreen wrapped round at the sides and blended smoothly into the body shape. No primitive 'aero' screen for a Lotus!

Practically all the feasible streamlining tricks were worked on the Lotus designs of this period. Most of these were not in themselves new, but Chapman ensured no detail was neglected. At this time, of course, not even Chapman had thought of using the low pressure under the car as a deliberate way of increasing tyre grip. It was taken into consideration mainly for ensuring there was no significant upward lift to cause instability.

Until the advent of cars using downward aerodynamic loads or ground effect, the only new ideas were on points of detail. Chapman was naturally to the fore in producing these, some of which were simply responses to changes in the rules to which the cars had to comply.

The Le Mans 24 hour race has always been a

RIGHT **The first purpose-built Lotus showroom at Hornsey, north London. The Elite just fits in. Outside is seen the 1957 Le Mans Index winning 750 cc Lotus Eleven.**

BELOW RIGHT **An excellent view of an early Eleven, showing off the overall smoothness of line. Note the wrap-round screen and metal tonneau.**

BELOW **Even in the early days, Lotus supplied cars to public figures. This Mk IX was supplied to James Dean – but only delivered shortly before his death. In 1987 this car was resold by Christie's for £27,500.**

LOTUS
1
TEAM LOTUS

11
11

ABOVE LEFT **This shot shows Colin Chapman testing the Lotus 19 at Snetterton. The dramatic skid marks show where he 'got it wrong' on an earlier lap. This sports racing car was a two seat version of the 18 Formula One car. It had its own tubular frame but most other parts were identical to those of the single seater.**

ABOVE **Chapman is seen accelerating from a hairpin bend, the nose of the 19 lifting under the thrust from the wheels. This movement can be cancelled by tricks of suspension geometry, as can braking 'dive'. Chapman had talked about these possibilities at the time, but the first experiments were not tried until some tests were done on a 21.**

LEFT **This view shows Chapman pressing quite hard. Slight understeer is evident, together with modest body roll. This was just about what he tried to achieve in a car at that time, giving a good compromise between peak performance and sufficient feel for the driver so that he could make good use of it.**

great challenge to the sports car makers of the world and it was not long before Lotus aspired to join the contest. 1955 saw the first entry by the Lotus team (already known as Team Lotus) in this great race. The car entered was a Mk IX on which disc brakes had been installed with help from Girling. The car had also been equipped with 'real' headlamps, streamlined behind perspex covers in the front wings. It retained the retractable lights to supplement them, but these did not meet the approval of the scrutineers on the Tuesday before the race. Lotus were obliged to position them inside the cooling air intake, one on each side. This was not the last time the Lotus team was to have difficulties at the hands of the scrutineers for Le Mans.

Because of the prominence, and some might say individuality, of Le Mans, it was often the Le Mans rules which most spurred Chapman to use his ingenuity. It must be said that this does not mean cheat, but to explore ways to comply with rules so as to incur the least design penalty. At the time of that first entry the rules allowed a narrow wrap-round screen, and for the passenger seat to be

ABOVE **The Le Mans 24 Hour Race 1957: the Lotus Eleven team cars had adopted this windscreen form with the tonneau cover from top of screen to rear body. This system was patented by aerodynamicist Frank Costin. This car is the 744 cc entry which won the Index of Performance that year, much to the dismay of the Panhard team, who had had a monopoly of this part of the competition. This car was capable of nearly 120 mph with its tiny Coventry Climax engine.**

LEFT **Le Mans 1958: the new tonneau was further improved for the Lotus Elevens and 15s the following year by making it inflatable to stop high speed flapping. Jay Chamberlain is helping in the preparation of his car. He was later slightly hurt in a crash near the Dunlop bridge.**

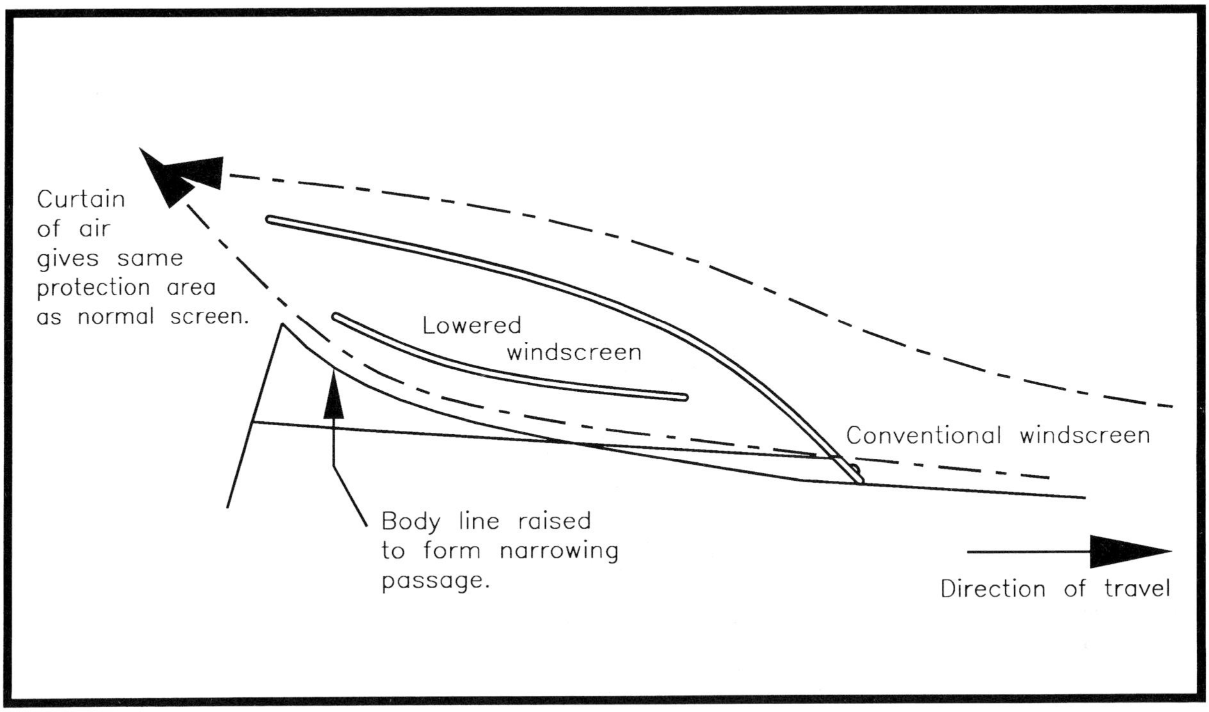

To achieve a lowering of the physical windscreen and give the driver a less obstructed field of view, Chapman came up with this venturi-style windscreen. It relied on producing a fast moving jet of air which itself functioned as a windscreen, deflecting the outer, slower moving layers upwards. The zone of protection for the driver thus reached as high with the lower screen as with a higher conventional one.

covered for the race with a metal tonneau. For 1956, full width windscreens were required to make the car less remote from a road-going one. Many of the sports racing cars of the time had a streamlined fairing behind the driver's head, intended to reduce drag and to save the driver from wind buffeting. The Lotus Eleven had been thus fitted during the 1956-7 season. By 1957, when Lotus were more seriously bidding for prizes, Chapman adopted a more radical approach.

The 'natural' result of the new rules would be that cars would have to revert to a conventional windscreen such as seen on a sports car of the time, an MG A for example. Chapman, however, was never one to take the obvious route unless it also happened to be the best and having a screen protruding above the smooth line of the body was by no means acceptable.

Lotus Le Mans cars for 1957 were fitted with a windscreen which curved over at its top to the horizontal and the bodywork behind the cockpit raised to this level. The tonneau cover was attached between these two points thus producing a shape complying with the new rule but offering scarcely any more drag than before. The fabric tonneau covers were prone to flapping during the race, so for the following year Chapman improved the idea by making them inflatable, like a small airbed so that the tubular structure added stiffness.

Fine-tuning airflow

When the Elite was developed in the late 1950s, Chapman was faced with the problem of how to put the rather large UK number plate on the front of such a beautifully streamlined car. The plate is of controlled dimensions and is required to be mounted vertically (many owners of cars fit or paint their numbers on sloping bonnets without falling foul of the law, but a manufacturer will not get away with it).

Study of the regulations revealed that the plate itself was defined only by its size and that there should be a black background for the white numbers (white and yellow plates came much later). Now white letters in front of a black duct leading to

a black radiator are clearly on a black background – so was created the mesh 'fly swatter' number plate. For the Elite the grille was made from a standard wire mesh, to which the moulded plastic numbers were riveted by the studs protruding from them. These never aligned correctly with the openings in the mesh, so when the Elan was designed several years later the grilles consisted of a tailor-made mesh where the spacings horizontally and vertically were different, but matched the spacings of the mounting studs of the numbers.

Chapman also came up with something original in racing car windscreens in the early 1960s. In the conventional racing car of the time a small windscreen was fitted ahead of the driver. The screen was transparent, but drivers had a better view if they were not actually looking through it.

For this reason the line of sight would normally be just above the top of the screen, relying on the upward airflow it caused to send the air over the driver's head and protect him from the wind of his passage. To reduce further the height of the screen would have been to have lost this benefit – until Chapman created the airjet screen, introduced on the Mk 25 in 1962/3. This worked on the venturi principle. A venturi is a deliberate narrowing in a passage through which the air flows. Its effect is that the air, pushed through the narrow space, 'hurries' to maintain the overall flow. In general terms this has two results, one is the increase in speed of the air and the other is a reduction in its pressure. In the airjet screen the increased speed is used to project the air steeply upward much more effectively than a simple screen just acting as a deflector. The upward flowing air then in turn

This close-up of Graham Hill's type 49 shows an actual example of the venturi-style windscreen.

deflects the oncoming air. The screen can thus be lower than normal, taking it further from the driver's line of sight.

It is not commonly realised that the tyres of a car are aerodynamic objects just as much as the body is. In fact, because part of the surface of the tyre is travelling forward at twice the speed of the car, they can be of very significant effect. With the tyre sizes in use since the 1970s the wheel almost resembles a roller rather than a disk, so this effect is even stronger.

The particular aspect that causes problems for the racing car designer (and driver) is that in the same way that a spinning tennis ball will veer in its path due to the aerodynamic side force generated by its spin, the spinning wheel generates a force which tends to lift it. One can picture the wheel riding over the air as it meets it.

It was these and other aerodynamic considerations that started the chain of events leading to perhaps the most celebrated occasion when Colin Chapman's ingenuity met with unexpected resistance from the motor racing establishment – the 'dual chassis' car.

Before this brilliant concept came along, however, the course of racing car design was to be changed by its predecessor, the ground effect car.

Ground effect

Looking at Formula One cars over the years one could always see under the cars as they went by. But with the introduction of the Lotus 78 'Wing Car' this was no longer possible. In the interests of lowering the centre of gravity, racing cars were gradually being developed to have less and less ground clearance, and by 1974 ground clearances of around two inches were normal. Cars had developed wings to hold them down and the idea was crystallised by Chapman when he saw the whole car as a wing, lifting downwards. To get the maximum performance from this wing skirts were incorporated, which kept control of the air flow under the car by preventing its escape and channelling it as required.

Once this step had been taken, the true ground effect car became inevitable. This idea moved on from using the skirts to optimise the wing function, to using them to create a suction area under the car. The wing car works away from the road surface – but better very close to it thanks to the skirts. The ground effect car only works because of its interaction with the road surface. It was the potential performance that was now available that led at last to the twin chassis layout.

With his analytical approach Chapman recognised was that there had evolved two distinct functional requirements in a racing car chassis. These were, on the one hand, to serve as a vehicle to carry the driver and some of the equipment with an acceptable degree of comfort and, on the other hand, a means to provide the maximum of aerodynamic downthrust to the tyres. These two functions bring in their train technical needs which are diametrically opposed. The downthrust requirement demands stiff springing and the need to use ground effect to generate it demands short suspension movements – these two are in any case consequences each of the other. The ride quality requirement demands soft suspension of generous travel.

It is quite evident to the observer of modern racing that factors providing the ride comfort have become subservient to those producing high cornering ability. The physical rigours of the racing driver's lot are again as hard as they were at the beginning of the sport, if not considerably more physically punishing.

The two functions were separated for the first time right at the beginning of the exploitation of downthrust. Some of the earliest wing systems were attached directly to the suspension members, thus putting the thrust where it was required. Meanwhile, the chassis itself was free to provide the level of ride the drivers had enjoyed in the preceding period.

One should not, however, assume that providing a good ride for the driver was simply a matter of keeping him cosy. A driver who is not subject to constant vibration will perform better and for longer, and can have a better command of the car – even if only because his feet stay on the pedals.

In the ten years from about 1955 there was a revolution in the ride quality of racing cars as suspension designers opted for lower unsprung weight in order to improve roadholding. This led to the use of softer suspension and hence a more comfortable ride. The benefits brought by the increased driver comfort and performance were a fortunate bonus.

Chapman was among the leaders in lowering spring stiffness along with the reduction of unsprung weight. He aimed for little damping from the shock absorbers on bump movement (so as not to transfer road shocks to the chassis) and heavy damping on rebound (to limit the resulting bounce action). These damper settings were combined with the softest springing that would ensure sufficiently fast recovery after a bump. In fact there were often times during testing where the delicate

balance was overstepped and a car would not be restored quickly enough after one bump before meeting the next. This phenomenon was known as 'ratchetting down' and it was disconcerting for the driver as the car would gradually ride lower and lower at high speed, particularly in a corner.

It was against this background that designers first used wings to improve cornering. It was natural, then, that those aiming for theoretical perfection would attach the wings to the suspension and not the body.

LEFT **In the South African Grand Prix of 4 March 1969 the suspension mounted wings are to be seen everywhere. Tall struts carry the wings above most of the turbulence created by the movement of the car. This time it was Jackie Stewart who won, in a Matra Ford.**

BELOW **The front wing attachment of the 49b. The cable controlling the angle of the wing can be seen. Because such attachments were difficult to make reliable and were easily damaged they were a potentially dangerous solution.**

It proved impractical to design such an installation to be reliable or accident proof. If a driver corners at a speed such that only the downthrust from the wings gives him enough grip to maintain control, then he really has no chance if the wing has been damaged or knocked off. Thus this type of wing attachment was banned and future wings had to be body mounted.

Once the wings became body mounted this inevitably led to the development of harder suspension (to transmit the downthrust to the wheels) and perhaps to ground effect – reducing suspension travel to preserve *some* ground clearance. This in turn meant a harder ride, driver fatigue, body belts etc., and perhaps led to the shortening of the Grand Prix race distance.

The point was being reached where the performance of the car/driver combination was limited by the fact that the driver could not give his best because the car did not provide an environment in which this was possible. The Lotus driver Stephen South, after testing the ground effect Lotus T80 at the French Paul Ricard circuit reported: '... so much vibration that I have difficulty in keeping my feet on the pedals and my backside's pretty bruised', and this was on a circuit known for its good surface. Chapman saw that the way to break out of this vicious circle was again to separate the purely 'chassis' function from the aerodynamic function.

One way of visualising what he did is to take one of the earliest wing-equipped cars with their wings mounted to the suspension. Next, imagine a framework, separate from the normal body, linking the front and rear wings into one efficient unit. This unit is not allowed to be direct-mounted to the suspension members, but has to be sprung.

Thus, in order to stay within the regulations, Chapman had to make a substantial outer 'body' able to ride on the suspension members and to apply the downward aerodynamic force to them. Within this he conceived another unit, the chassis proper, in which most of the functional units of the car were carried, along with the driver.

The outer unit was of ladder form, having side members of lightweight materials and steel cross bracing. This assembly rode on a suspension connected to the same main suspension linkage used by the inner chassis. The download produced by the aerodynamic effects was reacted as near to the wheels as possible without rendering impossible the provision of a good ride quality for the functional 'driver capsule'. Lotus likened the principle to that commonly used on heavy lorries of a sprung

5
LOTUS

STP
LOTUS

cab. Commercial vehicles need hard springing because of the loads they carry, so a separate sprung compartment is provided, not forgetting that there is often a suspended seat as well.

Chapman's aim was to improve the performance of the racing car by making a design which met the functional requirements as squarely as possible, allowing for all the diverse and conflicting priorities. It is well known that it is easier to find reasons for not doing something than it is to get on and do it. This is what happened when the new concept car was shown to the world.

The Formula 1 organisers and some constructors interpreted Chapman's design as 'driving between the rules' and banned it. How different would be the cars of today had they had the vision to see that it represented possibly the only avenue of development which could take the cars on to new levels of road holding performance without further taxing the driver. Chapman's scheme would not only have improved car performance but could have saved drivers from back problems in later life.

Tension rose during the design of the definitive car (the 88) following the reaction to its prototype (the 86) of the governing body FISA. Despite the commitment made to the constructors to leave the formula rules unchanged for a minimum period of two years, FISA introduced a new rule which effectively excluded 86. The new rule, which clearly added a new constraint to the formula, was passed off as a 'clarification'.

It was perhaps because of the tension caused by all the various arguments over the 86 that Chapman allowed himself to include some very light spring units in the attachments of the outer chassis unit, when very stiff units might (only might) have made the car's eventual acceptance more likely.

This is because the light spring units collapsed under the aerodynamic loads generated, whereas the very stiff springing already in use on Formula One cars would not have done. The ground clearance of the aerodynamic part of the chassis was therefore demonstrably variable. Other constructors, in fact, used some quite elaborate systems to reduce ground clearance once the car was on the circuit and away from the scrutineers – which is

LEFT **Jim Clark on opposite lock on his way to winning the British Grand Prix of 1967 in a Lotus 49 with no aerodynamic aids.**

BELOW LEFT **The first of the wedge noses was seen on the Lotus 56 turbine car for Indianapolis in 1968. Dick Scammell is helping Graham Hill.**

BELOW **Lotus continued to make club racing cars for sale until this, the 61 Formula Ford of 1969/70.**

what this approach achieved. They, however, were not operating revolutionary cars and therefore got away with it.

Peter Wright remembers Chapman's response once it was clear the car was not going to be accepted: 'Then what Chapman was fighting for was the right to innovate. He wasn't fighting for the 88 itself, he was fighting for the right try new ideas; in every court we went to, the legal technical specialists found the car legal. It was the political side that found it illegal. It was a very exciting time and I got to know a lot of lawyers very well. The issue was not really a technical one.

'I think there could be a case for saying it did not really meet the spirit of the rules but it certainly met the letter of the rules. The problem lay with the politics of the rules. Chapman said motor racing was all about trying ten different ideas to get one that worked; all he wanted was the right to try those ten ideas. He said if the rules were going so that you could not innovate then he was not really interested. This was when we went into microlights'.

In any field of competition of a technological nature the contestants must be sure that their products are not only capable of competing successfully, but also capable of further improvement to allow scope for reacting to their rivals' actions. Companies making a product must not be left behind by their competitors, or they may never catch up. They must know when it is useless to try to improve something that has little room for improvement, and thus when to seek new solutions altogether.

ABOVE **This 63 of 1969 shows an intermediate configuration of winglets and rear wing. Note also the wedge form.**

ABOVE RIGHT **The 76 of 1974 was an attempt to build a lighter version of the then ageing 72. The car had a very disappointing Grand Prix career.**

RIGHT **The Lotus 72 was highly successful and shows the wedge-shaped nose form first seen on the 56. This type won its first race in 1970 at Zandvoort and its last (72E) at Monza in the 1974 Italian Grand Prix. This photograph was taken during 1972.**

For example, ship builders had already developed the sailing ship to a highly efficient state when the first steam ships became available, thus the ships were close to their technological limits.

John Player Special
John Player Special
John Player Special
1
DUCKHAMS
DUCKHAMS
1

LUCKY STRIKE
1
LUCKY STRIKE RACING
SASOL
SASOL

John Player Special
DUCKHAMS
DUCKHAMS
1

John Player Special
John Player Special
5
Valvoline
Valvoline
GOODYEAR

LEFT **A Lotus 72 on opposite lock. This version was in use in 1974. Notice that some ground clearance is visible.**

BELOW LEFT **Lotus' entry for 1977 was the 78 wing car. This was a result of Chapman's far-sighted study of 1975, handed to his team just *before* the release of its predecessor the 77. Chapman realised they were behind all the opposition and so sat down to work out a new approach. Previous cars had wings added to passive bodies to thrust them down. For this one the body *is* the wing. No daylight can be seen under the car.**

BELOW **The last Lotus but one before the ground effect car was the 76. The car still has some ground clearance and light can be seen under it, though not much. It also possessed appreciable suspension movement.**

ABOVE **21 June 1970: paddock scene as Jochen Rindt's 72 is readied for the Dutch Grand Prix. He had made fastest practice time and won the race.**

ABOVE **From the rear a clear impression is given of the space under the ground effect car, bounded by the skirts, where the 'extra ingredient' makes itself felt.**

BELOW **From the front, the ground effect car appears to have no ground clearance at all, although the leading edge of the left hand skirt can be seen behind the wheel.**

TOP **The next step on from the 78 was to use the underside of the body to generate its own suction effect through the use of clever shaping of the underside into a venturi. In very simple terms the car scraped the air off the road at the front, leaving a vacuum which held the car down. The sides of the car were sealed by skirts.**

ABOVE **The operation of the ground effect car was so effective that the additional external wings were much smaller than before. The undershaping took advantage of the V form of the engine, thereby saving Cosworth from having to redesign it in flat form, which had been an imminent possibility.**

LOTUS
Lotus
TISSOT
WATCHES
MARTINI
Valvoline
NGK
MARTINI RACING
ESSEX
GOODYEAR
TISSOT
Valvoline
GOODYEAR

TISSOT
WATCHES
Valvoline
NGK
MARTINI RACING
ESSEX
GOODYEAR

ABOVE **The Type 79 succeeded very well and forced the opposition to re-think their own designs. If it had a fault it was that in producing the enormous downforce that gave it its dominance, it created a lot of drag. The 80 was designed to maintain the downforce while reducing drag, thus restoring the car's high speed performance. The car is seen crouching outside Ketteringham Hall in its original form without added wings.**

ABOVE LEFT **This view shows the Type 79 posed outside the main doors at the Lotus works at Hethel. Note one of the outlets for the under-car and cooling air above the side pod.**

LEFT **The Lotus 80. Instability problems proved so intractable that front and rear wings were added to try to salvage the project. The car never produced the results that were hoped for and the car was finally abandoned as something of a disaster.**

Some builders of sailing vessels, unable to accept the inevitable, continued to strive for performance improvements with which to match the speed and consistency offered by the steam driven rival products. Due to the high state of development already achieved, only minor improvement were possible, thus the exercise was fruitless and commercialy doomed.

Such manufacturers are in marketing terms 'defenders', trying to protect their market share against the 'attackers' who come out with new product concepts.

Motor racing, almost by definition, attracts people who are among nature's 'attackers'. Nevertheless, there are degrees of ability and vision within them and Chapman was certainly not one to wait to have to defend when it came to new ideas in racing car design. By this token the twin chassis Lotus cars were a step into a new technology, probably the only possible avenue open capable of allowing a significant improvement of performance

John Player Special
TISSOT
GOODYEAR
ESSEX

John Player Special
TISSOT
WATCHES
NGK
MICHELIN
11
ESSEX

ABOVE **De Angelis, Chapman and Nigel Mansell outside Ketteringham Hall.**

ABOVE LEFT **A Lotus 81 at Ketteringham Hall. Lotus' fortunes continued to be poor with the type 81. Again, it was a question of various changes being made to bring about performance gains, but these changes failing to deliver the goods.**

LEFT **Elio de Angelis poses with the new 87 at Ketteringham Hall.**

in one leap, as distinct from the progressive evolution of the established design. Such steps are like the opening of a gate from a well-worn field into a lush meadow waiting to be explored.

Aerodynamics and Lotus formula cars

It is interesting to trace through the development of the aerodynamic aspects of the formula racing car as exemplified by the various Lotus models.

The earliest model (the 12) was a cigar form, with the bodywork hugging the structure closely. Into the shape we can read the following aims in the design: minimising of frontal area and the reduction of drag of the resulting shape; attention paid to a clean nose shape to improve the penetration, complemented by an extended tail to minimise the drag induced as the air flows together after the passage of the car.

Closer examination shows attention to smooth air flow around the cockpit, with the screen wrapping right back behind the driver. The radiator sys-

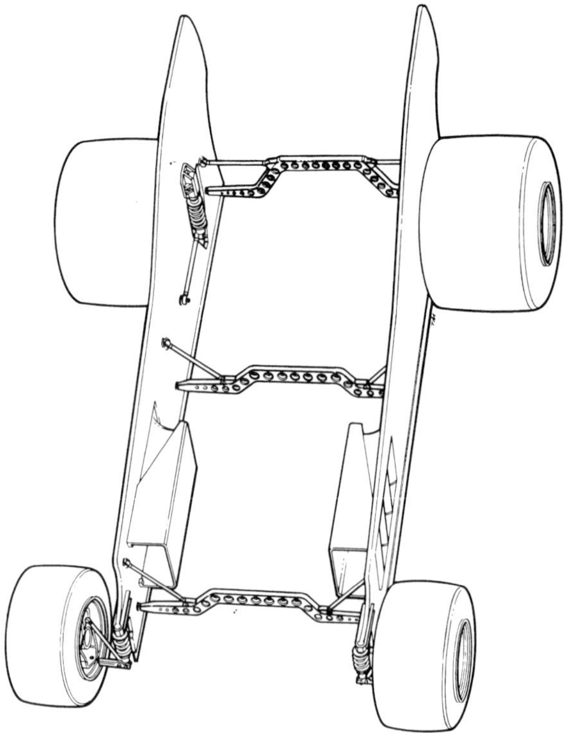

FIRST CHASSIS — LADDER TYPE
with independent outboard coil spring suspension system optimised to carry aerodynamic loads.

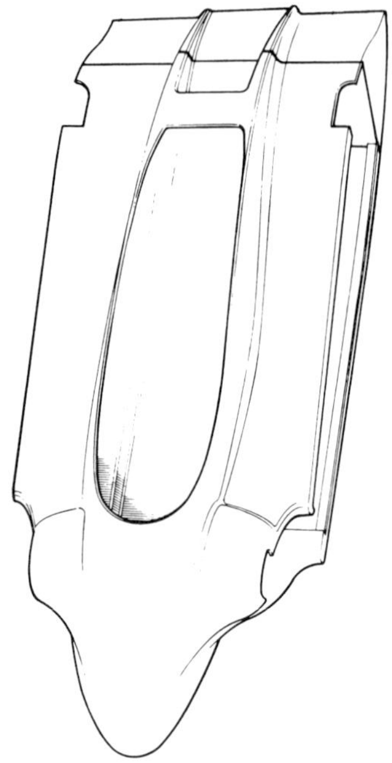

COACHWORK
making extensive use of carbon, Kevlar and Nomex honeycomb sandwich construction and all rigidly secured to the first chassis.

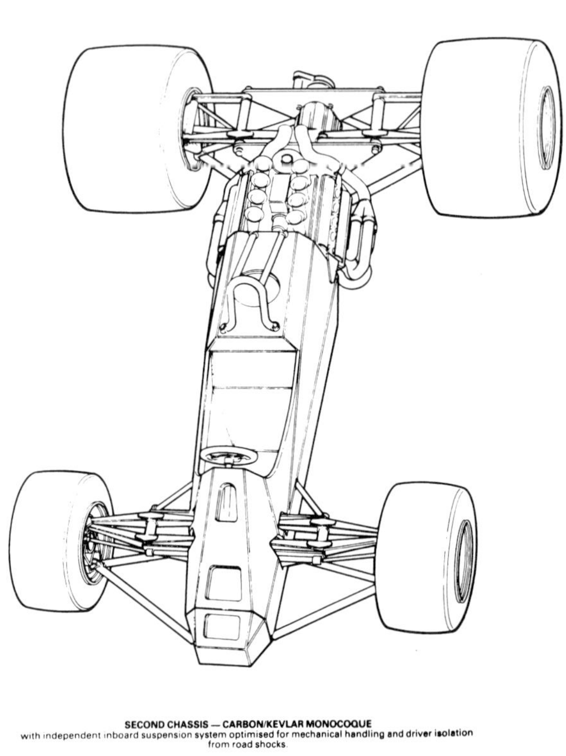

SECOND CHASSIS — CARBON/KEVLAR MONOCOQUE
with independent inboard suspension system optimised for mechanical handling and driver isolation from road shocks.

THE ESSEX LOTUS T88
a suspension development exercise.

LEFT **These four drawings by Tony Matthews of the 88 made part of the package released to the press at the time of the car's announcement. Bearing in mind the controversy over the first twin chassis car, Lotus were particularly at pains to make the best presentation. The wing shape of the aerodynamic outer structure is clearly shown, as is the almost fully functional inner capsule, carrying suspension and wheels. The separate suspension for the 'first' chassis is also shown.**

BELOW **The chassis jig for the 88 is shown in the Team Lotus machine shop. This is used to assemble the flat sections of panelling in lightweight composite material for bonding into a united whole.**

tem had minimal ducting.

As Chapman felt his way into the specialised field of the single seater racing car he developed his own theories about aerodynamics and began to prioritise the factors differently.

The Mk 16, which succeeded the Mk 12, had the now famous Vanwall form – elliptical cross section and flowing lines. Although by comparison with its contemporaries it was a miracle of compactness, it did not aim to achieve the minimum of frontal area, but a more perfect form. The driver had only a very small opening by which to enter the car, but once in had spare room around him – particularly for his arms and shoulders. The concept here was that there was no particular point in minimising the frontal area of the body itself since the wheels dic-

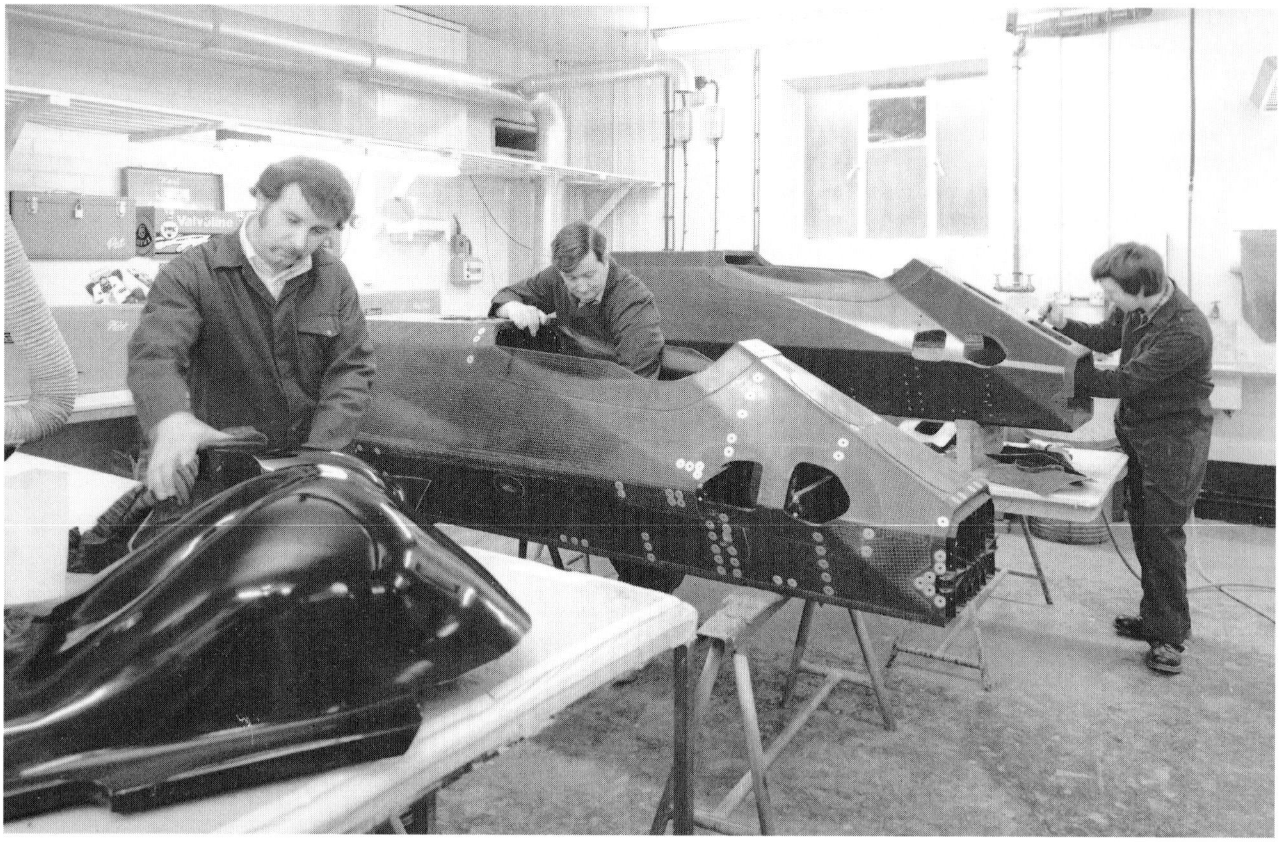

tated the effective frontal area of the overall result anyway. Steering lock was relatively restricted, following Chapman's precept of the time, and the body more or less filled the gap between the wheels. The front suspension units were outboard (although shrouded by the width of the body). The height and form of the tail showed consideration again for concern for directional stability.

Following the Mk 16, the next single-seater was of course the first rear-engine car, the Mk 18. This car reverted to the chassis-hugging body form of the Mk 12.

By now, Chapman had learned his lessons on steering lock, this being the first single seater to have the extended movement which characterised his designs for the next few years. Consequently there is a larger gap between the wheels and body, but the front suspension still has outboard spring units. There is no evident effort to balance lateral area to produce directional stability, although having the driver at the front instead of the engine ensures that the body profile can be lower ahead of him than behind.

Remembering that the 18 existed in Formula

Once the hull has been built up on the jig, this is dismantled to allow its removal. Two structures are seen being finished prior to assembly at Ketteringham Hall.

Junior form as well as in F1 (and F2) versions, the next car which shows the progress of Chapman's thinking was the 20, which only existed in FJ form. The 20 had the tight elliptical body form which carried through several subsequent models. It had a very slim nose and the driver sat lower, permitting a lower frontal area.

With the next F1 car, the 21, we are at the stage where Chapman's views on the importance of body streamlining, and the relevance of the rectangle enveloping the two front wheels in dictating effective frontal area, had changed completely. He had come to believe that a smooth airflow could be achieved as far back as rearward of the front wheels, but that the airflow was so confused by the time it reached the area of the rear wheels that little could be done. So the car was given inboard suspension at the front and the whole of the front designed to ensure the forward half of the car was as clean as possible.

The 24 and its monocoque twin, the 25, carried these ideas further. The nose section was even slimmer and the cleanliness of the front suspension was further improved by raising the steering track rod to the same level as the top wishbone, thus clearing the space between the wheel and body as completely as possible. Incidentally, this made it easier to obtain accurate steering geometry because the track rod and the upper suspension arm lay in the same plane and made a rectangle. In addition the suspension and steering pick-ups were rigidly mounted in close proximity on a strong box structure. This arrangement ensured accurate but 'zero Ackerman' steering. At this time Chapman believed that zero or negative Ackerman was desirable (see chapter 5).

Single seaters following the 25 maintained a remarkable consistency of body shape. These included the 29 and 34 which were built for the Indianapolis races of 1963 and 1964, the 27 monocoque Junior car, and the 33 F1 of 1965. Only with the 38 (Indianapolis, 1965 to 1967) and the 49 F1 of 1967 was there a visible change to the approach to aerodynamic form. These cars, of course, were designed by Len Terry. Naturally he was working to Chapman's schemes, but due to his very great experience and flair, he was allowed a much freer hand than had been the case before Terry joined the team.

The new shape had a slimmer and less rounded nose form resulting in an even cleaner entry shape. The space between body and front wheels remained clear. The glass fibre nose cone of the 49 was so slim that it had protruding fairings around the suspension top arm attachments.

It was on these cars that wings were seen for the first time in the form of small side winglets at the front and a simple wing mounted on very high struts on the rear suspension.

The purpose of the height of these mountings was to place the wing clear of the turbulent airflow close around the car so that the wing would be more effective and consistent. 1969 saw the development of the wings and the addition of high mounted front wings as well. Use was also made of a mechanism to 'feather' the wings for straight running, only putting them back into their high downward lift position for cornering. This was to reduce drag when the downthrust was not needed. Later, the 49, now without its suspension mounted wings (by then banned), appeared with body-mounted rear wings and the scene was set for the next radical steps already discussed.

Before these became visible however, more con-

This photograph of the finished car shows its unconventional lines.

ABOVE **For 1982, Team Lotus had the type 91. This car had water cooled brakes – as a ploy to reduce the weight by racing with the tank empty. This was the last of the true ground effect cars, since the following year the 'flat underside' rule would be introduced.**

BELOW **Once, when Chapman raced an Eleven in heavy rain, someone asked him about the handling – 'What is it like in the wet?'. Chapman replied 'Choppy round the ankles'. This cartoon from *Sports Car & Lotus Owner*, April 1959 is based on the same idea.**

tributing ideas were germinating. On the 56 gas turbine Indianapolis car of 1968 we see for the first time the flat wedge nose form, aimed at maximising penetration and down force.

This form, with or without the front cooling slit, was carried on several models right through to the 80 of 1979, thus spanning ten years of development. These ten years were perhaps aerodynamically the richest, bringing all the sophistication of the racing car as we know it today.

On the 56 the whole body was in the shape of a wedge. The shape, arrived at after wind tunnel experiments, can be thought of as a wedge which is trying to insert itself under the air above the road, thus being pressed downwards.

Chronologically, the next F1 car was the 63, in which the four wheel drive principle was tried, not altogether successfully. The body of this car showed the wedge form again, the nose tip having an opening for cooling, and it carried winglets and rear wings.

In 1971 a Formula One version of the 56 reappeared – the 56B, complete with gas turbine and four wheel drive. This car had the same basic wedge shape as the earlier car, but had the standard F1 aerodynamic aids of the time, consisting of a dual rear wing and a pair of front winglets.

Every subsequent F1 car from Lotus then possessed a recognisable wedge form up to perhaps the 76 of 1974.

'Perhaps' is the operative word here, because by this time the shape was very much narrower at the front, becoming more dart-like. In addition we were beginning to see the side radiator pods which became the key to the 'wing' and ground effect cars.

On the 78 'wing' car of 1977, the pods were extended to fill the space between front and rear wheels, maximising on the body surface available to act as a downward lifting wing.

This car cannot be said to be in the form of a wedge, although the form is still there in the main part of the hull. Due to its slab like side pods, the car, from some angles at least, resembled a chocolate bar on four wheels! By the time the 79 ground-effect car came along to dominate the 1978 season, the wedge had almost disappeared – except for the nose cone form, which still had some life in it.

For the Essex sponsored Type 81, the visual aspect was changed again, by mounting a transverse single wing across the nose, the last in the line before the attempts at the twin chassis design.

10 Engines and Gearboxes

COLIN CHAPMAN'S contribution to the design of engines and gearboxes was relatively small as a designer. His contribution as an entrepreneur, however, was much greater.

Through a combination of his own intuition and the natural way successful people attract others in their field, he 'collected' a number of engine experts. These included Keith Duckworth, later to become famous along with Mike Costin as the founders of Cosworth, the creators of many successful racing engines.

Steve Sanville was another such who joined Lotus in 1957 and stayed until 1969, becoming head of engine development. One of the first projects Steve worked on was the 5 speed Lotus transaxle – referred to as the 'queerbox'. In retrospect he regards the design as comical, while remaining aware of its very ingenious concept. In fact most of the concept work had been done either by Chapman or under his spur.

The principle was that the drive from the engine entered via what is the layshaft in a gearbox for a rear drive vehicle, drove the output shaft via the gears, and had the drive pinion on the output shaft. The adoption of this layout was no innovation in itself, since it was widely used in Volkswagen and Renault passenger cars. The innovation came in the way the rows of gears on each shaft were arranged to slide out from one end after the removal of a cover, and in the use of a motorcycle-style positive stop gear change mechanism. The gears were all made to be selected from a matched set and could be mounted on either shaft. A system like this means that the ratios are completely under the control of those setting up the car for a race and not in that of the gearbox designer. It is possible to have five ratios all the same, 'overdrive' on fifth gear, or even on all four.

When the early engineers were finding ways of making metallic gears work through a right angle, two forms were developed first, based on the experience of the wooden systems used till then. These were the worm drive, in which the axis of the driving gear (the worm) makes a tangent to the driven gear, and the bevel drive, in which the axis of the driving gear (now called the pinion) intersects that of the other. The worm drive is useful for very high ratio reduction drives and is smooth running and normally very rugged. Since the drive is transmitted by the thread of the worm wheel revolving against the teeth of the driven wheel there is a great deal of friction. This friction constitutes a loss of energy and creates heat. The pinion drive on the other hand can be designed to have very little friction and is therefore the more efficient of the two forms, but suffers from noise generation resulting from the successive tooth impacts.

Striving to produce smoother and quieter bevel drives, engineers next produced the spiral bevel – this can be thought of as a bevel gear which has been twisted, producing teeth which run around it in a spiral rather than lying along its axis. This results in a progressive engagement of the teeth and so is smoother and quieter.

Except with a transverse engine layout, vehicle final drive gears always work through a right angle drive, so early designers used both of the basic sorts of drive. Worm drives were often chosen for heavy vehicles for their smoothness and ruggedness. Designers seeking higher performance naturally chose the bevel drive, but soon found that they needed to lower the drive line somewhat to increase

the usable passenger space. To do this they adopted a drive in between the bevel and the worm. This system is called a hypoid bevel, or hypoid, drive. The hypoid drive is a compromise – it has more friction than a plain bevel drive but less than a worm drive, and is potentially quieter and more rugged than the bevel drive.

In order to suit the format Chapman had set down, the final drive gear was mounted below the centreline of the crown wheel – a hypoid drive – and it produced a significant amount of heat when transmitting the power of a racing engine.

Weaknesses in the original design abounded, not least in the lubrication system which has to carry away the heat produced as well as lubricating the whole mechanism. An early system used a jet of oil directed at the meshing point of the gears. The jet was too small and this resulted in more than one set of drive gears being left after test with no sign of any teeth. There were also problems in making the bearings of the gears on the input shaft reliable. In this design, the gears not selected were riding on the input shaft, driven by their mates on the output

ABOVE **At Hethel, the manufacture of the Twin Cam was brought under Lotus control at last. The last stages of assembly prior to test are shown.**

The Twin Cam was a close fit in the Europa. Tightly packed engine bays, however, are nothing new to Lotus.

The first of Lotus' own engines was the 907, 2 litre inclined 4 cylinder unit seen here in its original Jensen Healey form. Originally fitted in 1972, it produced 142 hp at 6,400 rpm. Performance was increased for the Elite version in 1974, to 150 hp, by increasing the compression ratio from 8:4 to 9:1 and a change of carburettors.

shaft, the speed of which was determined by the ratio in use. The bearing surfaces were just the steel of the shaft and gear, running together, a combination difficult to get right.

Problems also arose with the positive stop gear change system. Unlike a car system, where each gear has a separate position of the lever, this entailed the lever returning to the same position after each gear change, and therefore demanded a different driving technique.

It seems that many drivers found this difficult to adhere to in the heat of the race, and it has been said that, with the possible exception of Jim Clark, only John Surtees had the sensitivity to use the mechanism correctly. Surtees, of course, had been a multiple motorcycle world champion, which may account for this. The positive stop gear selector was replaced with a more conventional one before very long.

Overall, the gearbox has not gone down in history as a success, but it served Lotus reasonably well over several years and its essence saw service again in a re-vamped version in 1978. At the time the gear box project no doubt diverted much effort to its development which could have been put to better and different use. Lotus Formula One cars never had a trouble-free gearbox until they started using those made by the specialists.

Chapman's success in the world of racing gearbox design is less than glorious, but his contribution to engine design and exploitation is much more noteworthy.

The first 'Lotus' engine was to be a twin cam version of the Ford Consul engine of the 1950s. This engine, of 1703 cc was a four cylinder overhead valve design – and remembering that many Fords had sidevalve engines into this decade, it was fairly advanced for the engine of an ordinary saloon car.

Work started on the twin cam conversion for this engine while Lotus was still in its Hornsey prem-

A descendant of the 907 was this 2.2 litre version which went into production in 1984 in the SE Excel and later the Esprit. By now the capacity is larger and the compression ratio is at 10.9:1 and the power 180 hp. The engine continues in production at the time of writing.

ises, but the design never saw the light of day as the project was overtaken by events.

In 1958 the famous Ford Anglia model appeared with its own entirely new overhead valve engine. This was of 997 cc and of 'over square' dimensions. This is to say that the bore of the cylinders was greater than the stroke given by the crankshaft.

Over the years preceding this time car engines had been gradually recovering from the early taxation rules which favoured small bore engines. These taxation laws meant that cars attracted an annual tax based on their horsepower, calculated by the 'RAC Rating' formula. The formula assumed that the engine's power was more or less directly related to its bore size, which resulted in a generation of engines in popular cars whose cylinder bore was small in proportion to the stroke. This achieved higher actual power for the same taxation level. Manufacturers had so biased their designs that the development of engines was severely stunted by the artificial agency of the road tax system, long after the system itself had been done away with. Indeed the notion of a car being referred to as ten or twelve horsepower from the old rating system was still in use at that time.

Car makers recovered from this pressure only slowly, having developed their engine ranges under the old regime. Engine stroke/bore ratios came down progressively, but Ford's new entry of 1958 was the first to go so far.

The original version of the Anglia engine had three main bearings on its crankshaft and was quite light for its performance.

One of the performance limitations in an engine's design is the number of bearings the crankshaft runs in. This is because a well supported crankshaft will flex less under operating conditions and can therefore withstand a higher power output. With an in-line engine bearings can be placed between the cylinders and at each end, giving five bearings for a four cylinder engine. Thus the new

Ford, while good, had a limited potential for development.

It was, nonetheless, a very good basis for the engine for the popular Formula Junior racing, which it later came to dominate, and for the new small sports car in its early stages of development at Lotus – the Elan. The development of the Lotus Twin Cam engine is well documented elsewhere – notably in the book of that name by Miles Wilkins (Osprey). Our interest here is to look at the extent to which Chapman contributed, and how.

For a long time before this engine became a reality, Chapman had been clearly determined that one day Lotus would have their own range of engines.

Harry Munday was retained to carry out the design in detail, helped by Richard Ansdale. Harry had been chief designer at Coventry Climax, but was then technical editor at *The Autocar* magazine and was doing the Lotus job on the side, which meant that many of the design meetings were held in a pub near the Autocar offices. Chapman's influence in these early stages was very great and he was as usual a powerful motivating force. As the project got fully under way, however, he left the work more and more to his team. This way of working was typical of him – once he was satisfied that those helping him were working along the lines he wanted followed he would leave them to it.

It may have been generally true that Chapman

The Lotus 907 is seen here in its Series 2 Esprit form as made between 1977 and 1980. This is the 'low compression' version at 9.5:1 and produces 160 hp.

This view shows the version of the engine for the 1987 Automatic Excel. It is fitted with cruise control and continues in production at the time of writing.

did not interfere excessively in the work he had initiated, but on one occasion he did so unwittingly or perhaps thoughtlessly.

This occasion was when Steve Sanville had an engine under test. Following one run, Sanville removed some critical part and left the engine on the test rig while the part was examined. During lunch time Chapman brought round a visitor and, seeing the engine, apparently complete, decided to demonstrate it. The engine started alright, but within a matter of seconds stopped spectacularly. Steve still wonders what moved Chapman to such an irresponsible act.

Another example of how Chapman was sometimes inconsistent (we are all of us less than perfect) was in the solution to the persistent head gasket failures on the Twin Cam. The cylinder head of an engine is held down by a number of bolts to the block and forms the tops of the cylinders themselves. The cooling water passages in the head link up with those in the block across the join which must therefore be perfectly sealed against the high gas pressures and temperatures over the whole of the mating surface. The seal will only be as good as it should be if the surfaces of the two parts are flat and are stiff enough to remain so under all the stresses of operation, thus maintaining full pressure on the sealing gasket that is used.

The Twin Cam head showed a lack of stiffness between two of the combustion chambers resulting in local instantaneous reductions of the gasket clamping force at certain times during operation, provoking failure at this point. Sanville was looking at ways of strengthening the lower face of the head one day when Chapman looked by. 'Why don't you put in a rib?' he said. Steve thought it wiser not to remind him of another time when Chapman had scornfully said to him 'A tension rib in aluminium is not worth a light!', and went on to devise a way of putting in a strut going up to the cam box – a very stiff part of the head.

One of the most frustrating things that engineers have to learn to live with is the difference between the performance of a production example of their brainchild and that of the prototype made to test it. The troubles with the 'O' rings of the carburettor flanges were a case in point for Steve.

It was necessary to isolate the Weber carburettors from the vibration of the engine to avoid the problems this would cause – fuel frothing and poor mixture control for example.

This had been done on the racing cars for some time by fitting an 'O' ring in a groove machined in the face of the inlet manifold and then spring loading the carburettors up to them with spring washers on the bolts.

This worked well, but such machining would have been excessively costly on the integral inlet flanges of the production Twin Cam head. Steve therefore adopted a thin steel plate with 'O' rings moulded in position. Thus the faces either side could be flat, but the rings would be held accurately in place. A prototype set was made up and fitted to the Twin Cam left hand drive Anglia which was being used as a running test bed. This set worked perfectly and lasted until the engine was eventually scrapped. On the strength of this and other successful tests the design was used for production. However, early Twin Cam users suffered repeatedly from the disappearance of the rings into the engine. Why? Because under production conditions, when hundreds of parts are being turned out, it is often difficult to maintain the material quality and cleanliness during bonding from which the prototype benefitted.

There is no doubt that the Twin Cam Engine was highly successful overall. It did not satisfy Chapman's ambition to become an engine manufacturer, but it did act as a useful stepping stone towards this goal as it had created a team capable of tackling the task and had gained this team invaluable experience. The team was further enhanced

by the addition of Ron Burr, ex-Jaguar and Climax.

From this point Lotus started on the design of its first complete engine. Ron led the preliminary design studies to determine exactly what form the new engine should take. They undertook design studies for in-line and V-form four and six cylinder versions, in vertical, sloping and horizontal styles. The final design settled upon after the exhaustive study was a sloping straight four cylinder engine of 2 litres capacity. This could easily be developed into a V8 with the minimum of work, should the opportunity arise.

Once the key research had been done the work started in earnest on the detailed design of the complete engine.

At a time when the design was fairly well advanced for both the block and the head of the new engine, the team got wind of the fact that Vauxhall was bringing out an engine with a similar specification. So they made a visit to the Motor Show where the new Victor was being shown, to find out just how similar.

No-one could have guessed how much similarity there was between the new Vauxhall engine and the one still on the drawing boards at Lotus. Clearly the Vauxhall design was further advanced, since

The 907 engine spawned many later generations, right up to the 2.2 litre turbocharged seen here. Although not launched until nearly two years after Chapman's death, it is influenced by him. This was the first Lotus fuel injection engine, and with its turbo unit produced 215 hp, while meeting world-wide emission control standards. The engine was produced from 1984 to 1987.

here it was offered for sale. Apart from that, this engine had practically the same cylinder dimensions, the same crankshaft bearing sizing and positioning and the same slant angle. Almost the only technical difference visible to the astonished Lotus engineers was that the Vauxhall had its cylinders slightly closer together, making the engine a little shorter overall. It also differed in having a single camshaft cylinder head, but it was designed for overhead camshaft layout; unlike the Ford which had formed the basis for the Twin Cam.

Once over their surprise and understandable disappointment, the team realised that here was a potential opportunity to save themselves considerable development time and cost.

Vastly less would be involved in adapting the Lotus head design to suit the Vauxhall block than

in completing the design and test programme on the entire engine bottom end. No doubt the famous Chapman charm came into play in the initial negotiations with the larger company, and the outcome was a number of liaison meetings and Lotus being in receipt of the drawings and design information necessary. Thus Lotus completed what started out as its own engine as another twin cam conversion.

The fruits of the team's work went on to power the Jensen Healey and later the 'new' Lotus Elite, followed by many others. The descendants of these earlier engines are still in production at Hethel and have set Lotus up as an engine producer in its own right.

One further engine design was to be inspired by Chapman's genius. This was a very small engine destined for Chapman's microlight aircraft design and will be described in chapter 11.

It is perhaps an interesting insight into Chapman's own character, as well as into the way he viewed certain others, to recount a memory of Steve Sanville's when visiting Lamborghini with Chapman. The Lamborghini car company was

The ultimate version arising from the four cylinder Lotus engine was this V8 unit of 4 litre capacity. Clearly, many of the parts were common with the smaller unit and so both 4 litre and 4.4 litre versions were possible. Both types were made experimentally. Work started in 1979 and several test engines were made. The last of these was removed from an Esprit in 1991. No production engines were ever made.

started by Sig. Lamborghini once his tractor company had become sufficiently lucrative to allow it. It had been a long term ambition of Lamborghini's to go into car production. Entering the foyer with Steve on the occasion of this visit, Chapman was immediately taken with a small helicopter on show there and had a close look round it. At a suitable moment during their time with the firm's owner, Chapman asked him why it was there. Lamborghini replied that that was his next ambition; when the car firm could support it, he wanted to get into helicopter manufacture. Chapman had similar ambitions for diversification and took pleasure in this discovery.

11 *Other Fields*

AS WE HAVE ALREADY discovered, Chapman had always been interested in aircraft. From the Air Scouts and the Air Cadets he carried his love of aviation throughout his life and made considerable use of flying in the course of his career, first in a single engined Miles Messenger, later a Piper Comanche. Later (only partly because the company's insurers insisted) he moved to twin-engined aircraft and remained with these from then on. These included his last aircraft, a Cessna Chancellor turboprop appropriately registered G-PRIX. He also flew a Bell JetRanger helicopter at one period.

But to try to cover all his other interests in a reasonably chronological manner we must look back to about 1970 when Chapman acquired a financial interest in the boat-builders, Marauder Marine, to complement his personal, technical and aesthetic interests in boats. One of the people closest to the centre of many of the projects in which Chapman involved himself, was Colin Gething. He joined the company in January 1973 to work on the 14 metre (46 ft) Marauder. This boat had been exhibited at that same month's show in London, but was based on an established wooden construction design. Being the sort of person he was, Chapman could not be involved with something without generating ideas of his own and he set the task of re-designing it for manufacture in reinforced glass fibre, using the lightest possible construction.

The potential advantages of light weight for a boat are many; when a boat planes its weight is supported by lift generated by its passage through the water. As everybody knows, this causes the boat to lift out of the water and so reduces the drag caused by the movement of the hull. What is not so obvious is that the generation of this lift causes drag forces in relation to it, just as with the wing of an aircraft. The more the lift, the more the drag. The lift is equal to the weight of the boat; in other words, a lighter boat of a given size will generate less drag than a heavier one. Less drag means less power needed, so smaller engines can be used. These are lighter and need smaller fuel tanks, so the boat's weight will be even less. The designer of a boat can therefore gain performance and reduce operating costs by achieving a light structure. In any case, a lighter structure is usually cheaper as well because it uses less material. Alternatively he can use weight saved in the structure to allow for more lavish or sophisticated equipment for a given total weight.

Colin Gething set out to develop the techniques of vacuum assisted resin injection, with a particular view to using them for the mouldings required for the boat. He worked closely with Albert Adams, who had been Chapman's composite materials expert already for some ten years. The outcome, which was the successful moulding of the 14 metre hull in one piece, is difficult to put into context, but it should be remembered that this was the mid 1970s and the only glass fibre boats of comparable size were made by the older hand lay-up methods. To make such a component by an injection method was a very considerable achievement.

The use of vacuum to assist the inflow of the resin was already established as a principle and it was being used for any part that Lotus needed to mould in any quantity. Traditional injection methods rely on the generation of high pressures within a storage vessel or pump system, which requires moulds that can withstand the pressure. A mould for a large part sometimes has to withstand very great loads, since the loads are multiplied by four

each doubling of size. The moulds, needing to be more and more massively built the larger they are, become difficult and costly to make and are only commercially viable for long production runs.

Vacuum moulding saves many of the costs involved with the glass fibre process and means that good quality tooling can be used for parts with only modest quantity requirements. By simply drawing the air from the mould, atmospheric pressure acts on its outside to close it and also on the liquid resin flowing in from the reservoir. Something for nothing. 'God's gift'. It fitted well with Chapman's philosophy of engineering, and perhaps of life.

In working with Chapman, Gething was often asked to take Chapman's concepts and turn them into something workable. In particular, these concepts were sometimes presented in a very simplistic form, not taking account of the complexities of a three dimensional realization. This part was left to

Clive Chapman raced this boat, designed by his father, for about four years; first as a junior, then in the senior class. He is seen here competing in a race in 1980.

Gething who had perhaps sharper abilities at this sort of work. Chapman was thus, as ever, making best use of the skills of those around him, and involving them thoroughly in the work.

Work on the boats started from square one. Under Chapman's direction nothing was taken as correct, just because it happened to be an established practice. An example of this was the question of hull rigidity. Traditionally hulls are very solidly made, so as not to flex significantly. In order that a lower weight could be achieved, the Marauder hull was designed to possess controlled flexibility. The skin thicknesses used were around 6mm (¼ inch) as against the 12 mm adopted on more traditional

designs. The skin panels were bonded to a stiffening structure of members made from fibreglass made up rather like corrugated cardboard, but on a larger scale. The central vee form and the rear under surfaces of the hull which carry the weight during planing (and consequently most of the wave shocks) were nearer the normal thicknesses, but wherever possible a weight saving was realised.

Chapman encouraged the design team to err on the side of lightness. If the structure was to be the lightest possible, the only practical way open to them was to make something of extreme overall lightness, with the consequent risk that certain local points would not be strong enough. These weaknesses would be exposed by testing and removed by the addition of localised reinforcement. It should, perhaps be emphasized that all engineering is done in ignorance and often an allowance for this ignorance is made during the design. It is called a safety factor, and may be a question of multiplying the basic estimates of the strength a part needs by a factor which may be from 1.2 or so up to 4 or 5, depending on the design history and the consequences of a failure of the part. This may be called a safety factor, but in truth it is an ignorance factor, put there to try to allow for that which is not known. If absolutely every detail was known of the forces on the part and the way the part would resist this load pattern, no such factor would be necessary. Unfortunately no real-life situation allows us to know in advance what loads our design will be subject to during its future life.

Another fine racing shot of Clive Chapman's boat. As a junior he won the class championship.

If a boat is to be sold on the open market it needs to be designed along certain guidelines to make it acceptable to insurance companies (and thus to customers). The design work was therefore discussed at every stage with Lloyd's Register to ensure compliance with its rigid regulations. Many times the radical approach being made by the Lotus team led to lengthy discussion and explanation before the new idea was acceptable to Lloyd's. New ideas were not confined to the hull structure. The internal fittings and equipment were also reviewed with the same aims of saving weight and space. The mounting of the engines and the drive system received particular attention in order to reduce the noise and vibration levels in the cabin areas. The objective was to achieve a degree of refinement until then only found in a luxury car.

All these efforts towards weight saving resulted in a fully seaworthy boat of a total dry weight of only 7 tonnes, including the engines and all equipment. The original design, using conventional methods, had weighed all of 12 tonnes; a reduction of 40% had been achieved. The finished craft, fully enclosed with three cabins, could plane at 29 knots on a modest 420 hp.

Colin Chapman's contribution at this stage of the work was very real and intensive. He came up with technical ideas and wants lists for the team to work towards. He had little need for actively driving the work forward; his bubbling enthusiasm was a continual encouragement – to an already highly committed team, who seemed able to achieve more than they thought possible themselves.

After the Marauder, which was a boat adapted to the north European market, and consequently requiring weather protection and full interior accommodation, the team turned its interest to a boat to suit the needs of the Mediterranean user. As before, the starting point was a boat already existing in more conventional construction. The Mystere was the creation of an Italian company called Versil Craft and had the typical large open deck area and sunbathing spaces one would expect for the market. Lotus' plans for the boat included the conversion to their lightweight structure, together with the fitting of new engines located further aft than the originals, thus creating more internal space for additional accommodation. The Mystere had a length of 13 metres (42 ft).

Both the Mystere and the Marauder were well received by their respective markets and enjoyed commercial success, although it is rare to produce long production runs of boats in this class. Sales of these boats (and also of the Moonraker) were handled directly by a small team operating independently of any distributors, thus minimising sales overheads. The Moonraker had been Chapman's first boat venture when he bought the company making it, Moonraker Boats, from David Buxton. No major changes were made to it and Lotus relied on the boat's established reputation rather than trying to break any new ground.

Work started in the late 1970s towards Chapman's ambition of becoming a boat manufacturer in his own right with the inception first of the Mamba and later of the Mangusta. The first of these was a 10 metre (32 ft) high performance cruising boat which was produced and promoted briefly in a market that was heading for a slump. The Mangusta, breaking the moulding size barrier again, was to be 16.5 metres (54 ft) long and set to become the team's top of the range product. Sadly, by this time the market conditions had deteriorated to such an extent that the project was dropped before the design was completed. The boat companies were wound up in 1980.

Like every father, Colin Chapman ensured his son benefitted from his own knowledge and skills. When Clive Chapman was about thirteen, his father introduced him to junior power boat racing, perhaps hoping that this bug would bite him rather than the motor racing one. To start off with, Clive's boat was a production boat from one of the regular makers for this class. It was not long before his father's ideas came to the fore. Firstly he shortened the boat by simply cutting it in two and rejoining the parts. The race scrutineers were a bit puzzled at first, being sure it was different but not really seeing quite in what way.

The next step was for Chapman senior to design his own boat; naturally he could see many ways he

Clive Chapman with his father shortly before a race in 1981. In the senior class championship he was runner-up.

felt he could make improvements. Mainly he wanted to see a great weight saving, but in addition was developing his own version of the classic theories. He had bought several text books on hull design and fluid dynamics and, having soaked up what was in them, proceeded to build on it to produce a new boat for Clive. Coming from Colin Chapman one might have expected an advanced structure in composite materials, but no; he schemed out a boat in wooden construction. This was not a concealed touch of conservatism coming out, it was just that as the design would not be made in large enough numbers to justify the moulds, there was no better material for the advanced shapes he wanted than wood. The structure was mainly in ply, with reinforcing blocks of balsa, resin-bonded into a strong whole. With this boat and its regulation 15hp motor, 14-year-old Clive

This view of a hull, recently out of the mould and having the internal reinforcement added, gives a good impression of the size of the one piece moulding.

was able to compete with the best at a top speed of about 35mph. That year he won the National Junior Championship.

Two years later, Clive moved up to the senior class with the same basic boat, enlarged to take the bigger 60hp outboard now allowed. He was second in this class that year.

Since the craft was so light, it was perhaps to be expected that some problems might result. Indeed, Clive still remembers vividly several occasions of the boat taking off completely on a windy day, one time when the boat did a double roll in the air before coming down rightside up.

A number of boats nearing readiness for delivery.

This lightness led to a clash between father and son when Clive wanted to weight the front of the boat to reduce the tendency for taking off. Colin was absolutely opposed to this. For him it was right to distribute the weight you were obliged to carry to the best advantage; but to add weight to overcome a design problem was definitely not good. Clive, though, took the more pragmatic view that he could add a bag of sand in the front with a minimum of work and get a benefit for almost nothing. The boat, in any case, weighed only about 100lb.

Of baths and coffins

Here it may be appropriate to look back at some of the ancillary work that had been undertaken in the quest for successful moulding techniques and production processes. At an early stage of the process development in 1973, Chapman and his team looked around for some familiar objects that they could produce by the new methods. This had as its main objective proving that the production ideas would work, but if the guinea pig product could be sold as well, so much the better.

Thinking of things that have the same general form as a boat hull, but smaller, one comes up with baths, coffins, garden pools and so on. The team came up, no doubt, with such a list and the first two of these they actually made. For a while Chapman, and Colin Gething and his staff, were inspired by the idea of selling glass fibre coffins and a number of samples were made. A little further thought and one realises that if such a coffin is buried it will practically never decay, and that if it is cremated it

ABOVE **Lotus' largest boat project was the Mangusta which, due to the decline of the market, was never made.**

BELOW **A beautiful night-time shot of a Marauder quietly at rest.**

ABOVE **The smaller Mamba passes at high speed...** BELOW **... while a Moonraker cruises regally past.**

Moonraker

ABOVE **The JCL Marine Mirage – 'The Exotic One'.**

ABOVE LEFT **Chapman allowed his passion for boats free rein during his involvement. He is seen here at the helm of a Moonraker.**

LEFT **One of the Lotus boats was the Mystere, originally the creation of the Italian company Versil Craft.**

would make for some rather nasty pollution. In the light of this, the coffin project, although providing some useful lessons, was abandoned before too much embarrassment was experienced.

The next shape to receive attention was the bath-tub. This did not suffer from the same obvious drawbacks and in any case glass fibre and other plastics baths were already being produced by some of the established makers. The fact that this was already being done meant that on the one hand there should not be problems with customer acceptance, but on the other that there might be difficulties in achieving useful sales levels in the market place. As a way of supporting the work, Lotus managed to secure a contract from a firm called Bonsac Baths to develop a range of luxury quality and style. They made a number of moulds and several prototype baths, but the project was dropped.

The microlight project

As the market for luxury boats fell towards 1980, and with it the possibilities of fulfilment of Chapman's passion for them, Chapman set Gething and his team to work on another project very close to his heart. By 1980 public awareness of the microlight aircraft was increasing to a point where he felt he could make a useful input to this field with his

The design model of the finished Lotus Microlight craft is displayed against a photograph of some Californian scenery.

ideas. The microlight aircraft in its modern versions derives from the hang glider, which had given rise to a sport of widening interest and sophistication. The hang gliders of the time had their ancestry almost exclusively in the Rogallo wing, which used a structure of tubes and cables to support a triangular fabric wing. The pilot was suspended prone under the wing and his weight provided stability and control. There are still today many machines on the market using these same principles. Where they have developed is mainly in the wing form, which has become longer and narrower, achieving a much better performance.

A number of makers had taken the step of adding motor power to their hang gliders to produce an aircraft of more general appeal and usefulness. These too had the pilot exposed beneath them and still relied on the manipulation of the weight distribution of the craft for control.

It was the emergence of this type of craft that caused rules for their construction and use to be set down in the interests of safety. The regulations had to include a definition of the type of craft to which they applied and rules were made which allowed their operation without a full pilot's licence. Though there were some rules to define the class and to ensure safe operation, compared with the sophistication of the Formula 1 rules the field was enticingly clear for Chapman's fertile mind. He turned to microlight aircraft with relish, feeling that the absence of rules would be like mountain air after the city and would allow him the free rein he

As the model 'flies' overhead, we can appreciate its cleanliness of line.

The pilot looks around carefully as he passes low over farmland.

This side view of the model emphasises the slimness of the nose and the propeller angle.

Here the model is depicted taxiing.

always sought. There was one enticing prospect.

Commercially, what Chapman saw (and he was not alone) was that if an aircraft of conventional form with a closed cabin could be made light enough, it could fit into the microlight category and thus benefit from the freedom of use allowed, without the resistance to general sales of the conventional 'rags and rods' structure.

So it was that the microlight aircraft project was born. Chapman would involve more of his team as the work developed. Colin Gething was to use his expertise to develop the airframe structure itself, Ian Doble (now Vehicle Engineering Director) was to design the special engine for it and Patrick Peal (now General Manager of Business Development) was brought in on the project for his flying skills and aircraft knowledge.

Ian Doble joined Lotus in 1979 from Perkins Engines at Peterborough. His work at Lotus had begun with the 2.2 litre version of the twin cam engine for installation in the Elite and Esprit. One Wednesday during the summer shutdown of 1981, Doble was at Hethel when Chapman came along with a microlight engine and propeller wanting to know how much thrust they would produce. The engine was a KFM, an engine used regularly at that time on microlights. Almost casually Chapman said that he needed the information in time to leave for California on Saturday. Together with colleague Brian Angus, Doble improvised a mounting frame suspended from the hangar door guide rails. The mounting was designed to swing like a trapeze under the thrust of the propeller so that this could

The CFM Shadow, designed by David Cook, is still rated as the only microlight currently on the market to have a performance comparable with that of the Lotus design. Ironically perhaps, Cook hoped for a collaboration with Chapman, which never materialised.

be measured by a spring balance in a restraining cable. They could not find an anchor point for the cable so borrowed a works Bedford van and positioned it as they needed. They made up a throttle control by use of a cable from an Esprit (the longest to hand) by which they controlled the engine with some trepidation. On the Saturday morning they handed Chapman the results but were dismayed to be given another propeller for testing before his departure at 3 pm.

To simply re-run the test with another propeller would not have been much problem, but of course it was not as simple as that, since the new one was a pusher, not a tractor, as the first had been. A tractor propeller is the most normal type seen on propeller driven aeroplanes and it had been easy to measure its thrust by the restraining cable method. A pusher propeller, however, is a different matter since it pushes against the engine. They had therefore hastily to re-arrange much of their measurement system. Fortunately the anchor point – the van – was moveable. When they gave Chapman the results he looked pleased and asked them to have the engine and propeller packed up ready for him to take with him. This they just managed.

Soon after this visit Chapman involved Gething and Peal. The purchase was arranged of an Aerolite Eagle microlight to learn a bit about them. One early morning in summer Chapman organised for

This three quarter rear view shows the prototype under construction in the Mojave works of Bert Rutan. The grey area on the wing is the carbon fibre reinforcement for the main spar. The windscreen mould can be seen on the table in the background.

an instructor to teach his family and a number of others, including Patrick Peal, to fly it in the grounds of his house. Chapman, as always when tackling something new, bought books on the subject and for some while they were knee deep in books on aircraft structures, aerodynamics, propeller design, control systems and so on.

Later a more advanced microlight was bought, intended to serve as a testbed for the future engine. This was an Eipper Quicksilver MX2. Peal assembled it from the kit as delivered and carried out some proving flights.

Meanwhile Doble, under the guidance of Chapman and Tony Rudd, started work on the new engine. As with the first Lotus engine, they started with research into the different possible configurations open to them in terms of simplicity of design, performance potential, weight and tool cost. A four stroke system was chosen to avoid the difficulties of the more normal two stroke approach and some drawings were prepared. This first proposal to be put before Chapman was of fairly conventional concept, being designed to avoid as many pitfalls as possible by adherence to established principles. Chapman did not like it at all. He said it was a pudding of a design and looked like a shoe box. He said he wanted something with a bit of flair about it – a Lotus design. Rudd was so incensed by these remarks that he went home and worked over

ABOVE **Home at last. Some 9 months after the death of Colin Chapman his microlight aircraft is shown in public at an open day at the factory (August 1983). Patrick Peal climbs away on a demonstration flight.**

ABOVE LEFT **This front view shows the foreplane and its carbon fibre reinforcement and the wing cavities to accommodate the fuel tanks.**

LEFT **The engine installation of the prototype. At this stage a standard Rotax engine was used, the Lotus engine being still under development.**

the weekend to come up with something better. One can surmise that this was exactly what Chapman calculated upon.

The new design was a radical approach, but when shown it Chapman said 'I did not mean you to go to the opposite extreme and come up with a Swiss watch!'. However Rudd's reasoning convinced Chapman that the outline design was workable as well as having the elegance he sought. It was then worked up into a full design by Doble, still under the eyes of Rudd and Chapman.

Looking today at the design, it really has a typical Lotus elegance. The layout is flat, with two push-rod operated valves per cylinder and the camshaft at the top of the engine above the crankshaft. The shell of the engine consists of only three main parts. A cylinder casting, a front housing containing the gear drive, and a rear housing containing the oil pump. There was no separate crankcase or sump, and the inlet manifolding was mostly within the cylinder castings underneath. The heads were also integral with the cylinders and they were cast in a high grade of aluminium able to be used without the normal iron liners. Incredible simplicity was achieved along with an invaluable weight saving. Some cost was spared by adopting the valves and rockers from the Vauxhall Chevette.

The engine was to be made in two versions of 25 and 50 hp respectively. For the 25 hp version two cylinders would be bolted together, with four being used for the 50 hp version.

One of the main constraints of the design was that the required horsepower had to be produced at 2500 rpm, since this was the highest speed at which the propeller could work effectively. After some research into previous engine designs using the same idea, it was decided to drive the propeller from the camshaft, allowing a crankshaft speed of 5000 rpm to be used without separate reduction gearing. Naturally the camshaft gearing would need to be more robust to transmit the full engine power.

In order to learn as much as possible before starting on the airframe, Chapman and Gething looked very closely at design and construction techniques in the two areas most closely related. These were gliders – of which much expertise was concentrated in Germany – and the American aircraft designed for home-building.

In America, where individual freedom is of so much importance, the regulations are such that the construction of aircraft by individuals is more favoured than in any other country. Consequently there has grown up a wealth of skill and knowledge on how to design and build aircraft as cheaply, simply and as light as possible. The home manufacture of aircraft had reached the stage in America where numerous companies offered designs, components and complete kits to the private builder. Some of these companies were run by highly professional and innovative designers and it is far from unknown for the designers of the world's passenger aircraft to visit the annual homebuilders' convention at Oshkosh to pick up new ideas. One such improvement to graduate from the kit market to the public transport field is the addition of vertical 'winglets' that we are now seeing on the latest commercial airliners.

This does not mean that all the work was done for the Lotus team. The home-built aircraft in the

This development photo of the Lotus Microlight engine on the test bed shows a 4 cylinder (50 hp) version. Note the construction, using two identical castings meeting at the centre line. The propeller is shielded for the safety of the test personnel.

When fully installed the engine was enclosed in ducting to provide a controlled cooling airflow.

United States were at the time almost exclusively of a larger and heavier design than that which would comply with the microlight regulations, but they represented the best fund of knowledge available for lightweight powered aircraft. Of the most original designers at the time were Jim Bede, who was already offering a nearly 400 mph single seat jet aircraft of only 14 ft (4.25 metre) wingspan for home-build, and Bert Rutan, whose speciality was for very high efficiency craft able to achieve high speed on low power.

However, the Americans do not have a monopoly in advanced and effective light aircraft design. At one stage Chapman met Suffolk-based David Cook, the creator of the CFM Shadow range of microlight aircraft. In their conversation, which started more or less as a social chat between two creative engineers, Cook said he was working on a new design with objectives of 100 mph on 50 hp. From Chapman's cagey reaction it was evident to Cook that Lotus had set very similar targets. Cook suggested a collaboration. Unfortunately for Cook, and possibly for the whole project, Chapman preferred to stay with Rutan. Rutan had, by masterful presentation of his not inconsiderable skills, built a world-wide reputation. He did not want a collaboration with Lotus, but a sub-contract for the design and construction of the prototype. His charges for this were consistent with his reputation. Cook, however, not only would have made a smaller charge but was keen to collaborate. The fact that his works were fifty miles from Lotus and not 6,000 would also have allowed for a better working relationship. Had Chapman liked the idea of working with Cook, the project may not have stopped when he died.

Whatever other possibilities may have been open, Chapman decided to work with Rutan. The best structures were then being made in resin bonded glass fibre, reinforced with carbon fibre, and making extensive use of styrene foam. Rutan was familiar with these methods and the design benefits and constraints their use implied. He had shown himself to be a very creative and original thinker in terms of the aerodynamic concept of the airframe itself. Rutan specialised in the 'canard' layout, in which the stabilising surface (usually thought of as the tailplane) is ahead of the wing. He saw that this layout, while deviating from common practice, offered performance and handling advantages if properly executed – and let it not be forgotten that the Wright brothers had used this same layout in the first successful powered aircraft.

Rutan produced a range of designs of canard form including the Vari-eze and Long-eze of light weight and very clean form, which had quite astonishing performance for their engine power.

Remember that the notion of speed to an aircraft operator is quite different from that which applies to car drivers. If one aircraft design can carry two people at 100 mph on 100 hp and another can achieve 120 mph on the same power, then the second will be more economical to operate, as well as giving a saving in journey times. Speed in aircraft is therefore mostly to do with functional and commercial viability.

The scheme Rutan produced for Chapman was also of canard form. The Gething team's role was to look at the design from the point of view of production, using a combination of the methods familiar to Rutan, the best of those used in glider construction in Germany and the processes they themselves had already developed for car and boat use. Once again an analytical approach was taken to all aspects of the manufacture and design, in order to produce something capable of setting a new market trend in performance and cost within the microlight formula.

Rutan built the prototype according to his normal principles and it was consequently above the target weight by some 10%. However a full study had been made of the weight and the scope for its reduction, thanks to Gething's work. The team were confident that, when the production version was made, the overall weight would held within the 330lb (150 kg) limit for the microlight class. We are talking about a complete aircraft to carry two people and five gallons of fuel, which weighs only the same as two people. The craft was to have a

ABOVE **The head-on view of the prototype, with the cockpit open.**

ABOVE LEFT **For the test and appraisal flights in America, Patrick Peal went to see the aircraft at the Rutan works before accepting it. He took a number of views at the time. This one shows him seated in the machine.**

LEFT **The aircraft outside its build shop.**

RIGHT **This view shows that to enter the cockpit might demand a certain agility.**

cruising speed of 70 kts (80 mph, 130 kmh) on its 25 hp motor and in still air cover 60 miles on a gallon of motor fuel.

Sadly the message from Rutan that the prototype was ready to fly arrived at Hethel on the day that Chapman died, so he never had the satisfaction of knowing of its completion. It was brought to England in 1983 and the honour of demonstrating it at the Lotus Open Day of July that year went to Patrick Peal, who had learned to fly an Eagle microlight at Chapman's behest two years previously.

Furniture and other ventures

Another of Chapman's pet interests was furniture design and the first time he turned his hand to this was while the company was still at Cheshunt, in about 1963. With the input of Ron Hickman, later famous as the inventor of the Black & Decker WorkMate, a unit was conceived for sale for transit lounges. Seen from above this was almost circular in form and consisted of three seats arranged facing outwards with the flattish sections between them acting as shelves, completing the shape. The backs

Note the rear skid to protect the propeller.

This view shows the seat cushioning. Note that the leg support is longer than the back support.

Note the underslung pivot for the elevators. This results in an increase in effective area as they are lowered.

The hollow wing fairing provides elbow space for the occupants.

Patrick Peal took several internal shots for record purposes. Note in this view the simple instrument layout. Production examples could have had more sophistication. The central control stick can be used from either seat.

This close-up of the control column shows the rod linking it to the foreplane control surfaces.

ABOVE **Compare this flying shot with those of the model. Note that the nose wheel is retracted but the main wheels are fixed. This was a common feature on Rutan designs.**

LEFT **This is a show model of the two cylinder version prepared before the production design was finalised. This unit is now at Hethel.**

of the seats together made a pyramid form. The unit was made from a one piece glass fibre moulding and was mounted on a tubular frame. A person was taken on to handle the marketing of the product, together with some variants based on the same ideas. However, when the costs were worked out, it was found that for the product to sell at £36, Lotus would get £12, the distributor £12 and the final outlet another £12. The fact that Lotus could only realise the same income per seat as the middleman, out of which they had to buy the materials and make the product, to say nothing of amortizing

the development and tooling cost, brought the whole thing to a complete stop.

It is not surprising, therefore, that some 15 years later, when Colin Gething's team set to work on a new furniture product, the intention was for sales to be handled directly. This product was also a seat, but this time a luxury reclining chair, exploiting the Lotus structural principles, and beautifully finished with leather upholstery. The chair won a design award but came to nothing commercially.

Gething's people were not occupied completely by original ideas under development. They had other work to fill in and to bring in some much-needed revenue. A contract was gained from a supplier of window frames to supply glass fibre moulded front doors, and a number of these were delivered. Another bread-and-butter job was as a supplier of body mould tools to De Lorean cars in Northern Ireland.

Projects motivated by hopes of a successful diversification were not the only things that Colin Gething was called upon to develop. Several jobs came along which were in truth purely for Chapman's personal advantage. One of these was to design a retractable cover for the indoor swimming pool at East Carleton Manor, Chapman's home. On the occasion of the marriage of Chapman's daughter Jane, he wanted to have a dance floor for the reception which was to be held at the house. No suitable space was available until he hit upon the swimming pool. Gething was then called in to create a set of suitable covers to allow an easy change from pool to dance floor. The same construction was used as had been developed for boat decks – smooth upper and lower surfaces separated by a corrugated glass fibre member, forming a series of box beams. The vacuum process was used for the bonding together of the different layers. The sections of the cover moved on rubber-tyred wheels running in the side gutters of the pool and

LEFT **Here a half-body shell is lifted from the Vacuum Assisted Resin Injection Mould.**

BELOW LEFT **After curing of the new body the mould is separated by means of the four pillar hoist. Note the pointed guide pegs for alignment of the top and bottom parts of the mould.**

BELOW **The Lotus recliner. Colin Gething worked on this chair in 1973. It was awarded a prize for design, but was not commercially exploited.**

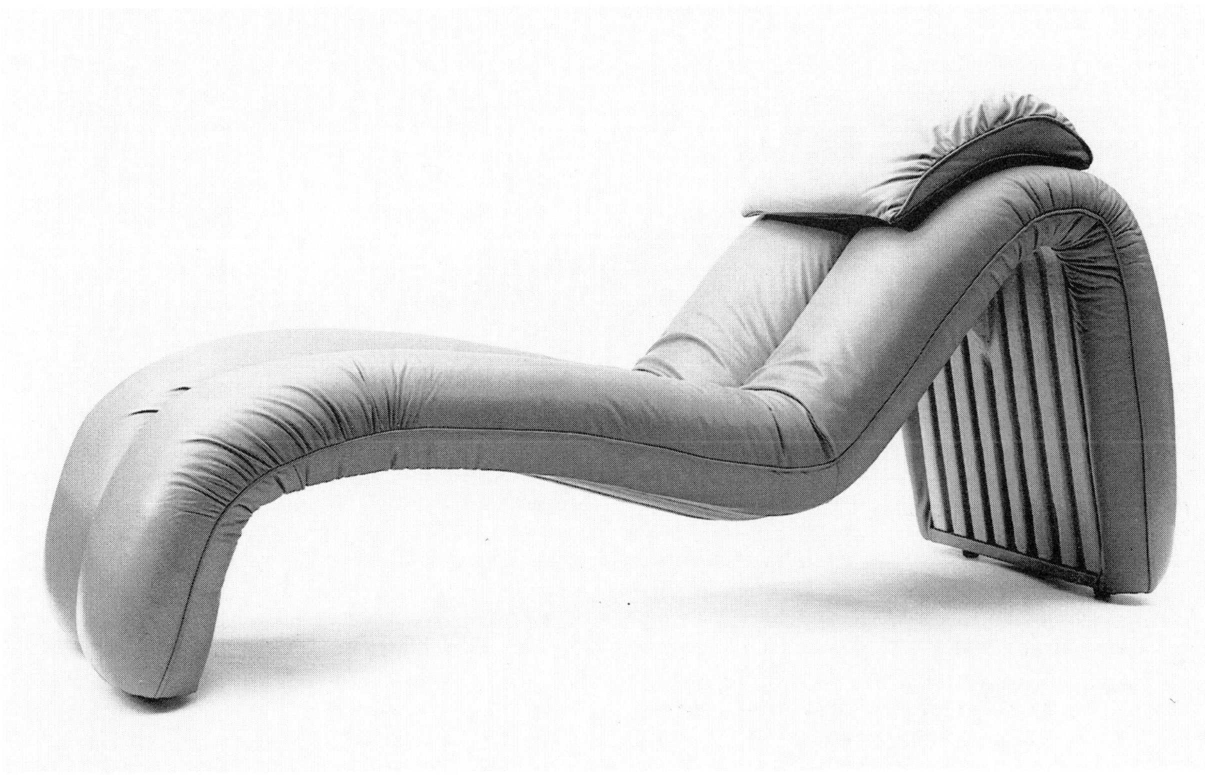

In this view a Lotus four cylinder engine can be seen in the foreground and a twenty-year-old Lotus Cortina in the background.

were electrically driven. They retracted under a false floor at one end when not required.

Colin Gething, in common with most who worked for him, remembers Chapman with great respect, affection and pride in having known him, coloured with a touch of frustration from his sometimes precipitate behaviour. When working at Ketteringham Hall, a few miles away, Gething was quite often taken by Chapman at short notice to the factory at Hethel to look at some problem or idea. On more than one of these occasions Chapman left Gething to investigate whatever it was and went off to deal with something else. This dealt with, he would then go off back to Ketteringham Hall, apparently forgetting about Gething and leaving him to get himself back.

This cavalier attitude to others showed itself many times, and at least twice in very similar circumstances. Gething remembers: 'It nearly happened to me in Italy also – we'd been to the Genoa boat show. Going back to his plane afterwards he went straight through Customs leaving me to sort the paperwork out, pay off the hire car etc. By the time I got through Customs he was already taxiing down the runway – so I had to run after him to get back home. That was a normal situation'.

Another long serving Lotus man still with them is Maurice Dowton who, with a colleague, having worked the usual excessive hours preparing a car for a showing in Paris, was invited by Chapman in return to fly with him and visit the show. Dowton arrived with Chapman at Le Bourget airport and travelled to the show with him. At the end of the day they were to meet at the stand for the return, but to Dowton's amazement they learnt that Chapman had already left with another visitor. As they spoke no French, had no French money and did not know where Le Bourget was, there followed some minutes of panic. They got themselves to the airfield, however, and finally made it clear they had to go to where the little aeroplanes take off. When they found Chapman in a departure area with his guest, his reaction on seeing them arrive was to turn to the guest and say 'There you are. I told you my lads would do it!'.

On the death of Colin Chapman on 16th

The prototype is shown in a marquee at the Lotus open day in 1983. A British registration has been added, although the US one remains.

December 1982 the company's board was naturally obliged to take stock and to clarify the route for the future. One such piece of clarification was to announce that henceforward the company would consider itself to be within the motor industry, and that the various subsidiary projects should cease. Without the lead from 'the guv'nor', this was no doubt about all that could be done at the time. It is interesting to imagine, though, the technical achievements they could have led to.

This view shows the first item of furniture from Lotus. It dates from about 1963 and was conceived as a unit for waiting areas such as airport lounges or station concourses. It was dropped when it was realised that it would not bring a reasonable return. The design was by Ron Hickman, later to make his name as the inventor of the Black & Decker WorkMate, and it was to be the first of a series to sell under the name Line Lotus.

12 Conclusion

THAT COLIN CHAPMAN was a genius has been well recognised for many years. The purpose of this book is not to prove this because such proof is not necessary. The object has been, rather, to shed a little light on the nature of that genius and the way it has affected those in the fields in which he worked.

There are naturally people who fell foul of him, or did not like him for one reason or another, but without exception, all those consulted gave their opinions enthusiastically and not one failed to acknowledge his abilities.

In his later years Chapman was increasingly dogged by one thing or another, be it the twin chassis car or his sometimes stifled passion for freedom to innovate. This freedom he sought again through microlight aircraft, but there, having chosen Bert Rutan as a partner, he felt thwarted because Rutan did not want to share his expertise in the way Chapman would have liked.

It is, perhaps, one of the tragedies of Chapman's life that in some of the many areas he tackled, he achieved less than he had hoped. The boat projects, although valued by many, did not support a stable business for any length of time and several projects were stillborn. Similarly his forays into furniture never really got beyond a few dreams, exciting though these were for those involved.

As for the microlight aircraft, it is impossible to say what would have become of it. It looks fairly sure, though, that development costs would have been far higher than he at first hoped, and that the final performance of the finished craft might well not have been as good as estimated. The increases of development cost are not just predicted because such costs always do rise (which is unfortunately

Colin Chapman is seen with Denis Jenkinson. Jenkinson was always a fan of Chapman, and often talked to him to get background for his articles in *Motor Sport* magazine.

Jenkinson discusses a point on suspension with Colin Chapman. The two first met when Chapman was starting to build a 750 Formula Car Club for the first time. He went along to a meeting of the 750 Motor Club to pick the brains of those already in the know. After one of these early meetings, Jenkinson commented when Chapman had left: 'I don't know who that chap is but he learns jolly quickly'.

true), but because of Rutan's perhaps understandable reluctance to 'teach Chapman his trade' while exercising it. This would have meant that the development and productionisation phases of the work would have necessarily taken more effort than expected and consequently cost more. These costs would obviously increase the minimum selling price of the finished product and so reduce its market potential. Things were further complicated by the wranglings of other participants, some of whom wanted more of a share than Chapman wanted to give. Perhaps if he had worked with CFM as David Cook the founder had hoped, some of these difficulties would not have existed, or have been less in impact.

It is interesting to speculate on the possible future of the microlight engine project, had Chapman lived to see it through. The engine concept was elegant and sound, and it could have been developed into a successful unit. Because it was potentially saleable in its own right, its development costs would have been spread over a much larger number than those of the aircraft itself.

Posing with your aircraft and sportscar. Notice the aircraft's registration.

LJH
792
JUST
MARRIED

PPW 306R

ABOVE **Driving your own Formula 1 car. Colin Chapman wearing a suit, photographed in March 1967, in a 49. The car shows its descendency from the original monocoque 25 – and thus from the 20, although the form has slimmed considerably in the six or so years in between. A framed picture similar to this decorates the foyer of the Team Lotus headquarters, Ketteringham Hall.**

Lotus Cars in films – the wedding car: The Mk II, which Colin Chapman built in 1950 is seen in Brothers in Law. (From the cover of *Sports Car & Lotus Owner*, June 1957.)

LEFT **Lotus Cars in films – the sea-monster: An Esprit is seen emerging from the sea in a James Bond film of the 1970s.**

There were and still are only a few suppliers of small aircraft engines, and of these most had two stroke units adapted for aircraft use. The Lotus engine was a four stroke conceived from the outset to meet the demands of flying. The simple modular construction should have simplified maintenance and the problems of providing a service. There is, however, one large unknown area and that is the question of the obtaining of the necessary approvals from the national and international bodies responsible for aircraft safety. While certainly not out of reach for the Lotus team, these approvals would constitute a very costly addition to the overall project development costs – again affecting selling price. It is pleasing to speculate that once through these tests the unit might have taken its place as a British product in this specialised market.

It was in the later years that the first whiff of fraud arose over the alleged missing millions of the De Lorean project. It is not relevant to this book to try to investigate this area, but many express suspicion over the degree of Chapman's involvement. Most wonder why a man so admired by the world and wealthy by any standard should be tempted by the possibilities in view. Those who believe him guilty answer that it was not the extra wealth that would have attracted him but the challenge presented to his ingenuity. Few would argue that he was not ready to respond to such a challenge, but that does not mean it is proved.

It is sad that, with a man so uniquely talented,

there arose some rumours of scandal. It is not intended to go into these issues, but since the rumours have more recently become more substantial, mention must be made of some of the known facts. It would appear that he was not blameless in the De Lorean affair, being involved in extracting several millions of taxpayers' money for his own use. Chapman's long-standing financial guide Fred Bushell was gaoled recently for his part in the affair, clearly implicating Chapman as another guilty party. It is perhaps surprising that Chapman, a man who 'had everything' would find it necessary to go in for that sort of thing; but two explanations present themselves: firstly, that Chapman (and perhaps Bushell too) was put in a position where to carry through the swindle was easy and to back out would have destroyed the relationships he had with others that allowed his legitimate aims to be pursued. Secondly, that as

ABOVE **Showing your brochure to royalty.**

BELOW **Explaining your cars to royalty on your own Motor Show stand.**

Chapman always gained satisfaction from finding ways of achieving the impossible, perhaps it was the pure challenge that drove him to it. The third possibility, that he was a fundamentally dishonest person is harder to believe – would it not have shown earlier in life? Explanations do not, of course, excuse.

There are even those who say that Chapman's death was faked so that he could escape to some distant haven. This idea is so attractive to some who have a terrific faith in his abilities that they see it as a way he might have got back at the authorities who often thwarted him. However, those involved in this deal, if it happened, were not those of the racing world, where most of his frustrations arose.

The Grand Prix of Norwich. The number of people who turned out to see the team go by can be judged from this view of St Stephen Street.

In meeting many people who knew him well, who knew the family over many years, there has, however, been nothing to give this author the idea that such might have been the reality. Perhaps a work of fiction might have ended this way, but not this book.

Colin Chapman has been dead for more than a decade now. Since then, his beloved Team Lotus has been through some difficult times. The Team won the Austrian Grand Prix in August 1982 with a type 91, before Chapman's death; however it was not until the Portuguese Grand Prix of April 1985 with the type 97 that they won again. Since that time their fortunes have waned considerably, with difficulties in finding sponsorship. The 1992 season saw, much to the relief of the Lotus people and to the satisfaction of many followers of motor sport (not necessarily Lotus fans), a steady improvement in their cars' performance thanks to a

ABOVE **In 1978 Lotus won the Constructors' Championship, beating Ferrari by a margin of 28 points. One result of this triumph was that the team was given the Freedom of the City of Norwich. The team paraded through the city led here by Andretti in his Type 79, here seen coming along Queen's Road, which is now a dual carriageway.**

RIGHT **For many the highlight was waving to the drivers who passed so close. Andretti is seen just after passing St Giles' Church in St Giles' Street, Norwich, heading for City Hall.**

new sponsorship deal from the British oil company, Castrol, and Johnny Herbert and Alessandro Zanardi carried the fight into 1993. Chapman's legacy lives on.

ABOVE **The seal was set on the day by the Civic reception at which Chapman is seen in his speech of thanks. Behind him are the Lord Mayor of Norwich, Dr J. T. English, and Mario Andretti.**

BELOW **Colin Chapman drove himself in the cavalcade in an Elite. It is interesting how diverse are those gathered at Chapelfield showing their support; not all are archetypal racing enthusiasts.**

ABOVE **Signing autographs. This was the scene on the occasion of Lotus being given the Freedom of Norwich.**

ABOVE LEFT **This aerial view of the factory at Hethel shows the extent to which it has been developed. The site was a World War Two airfield and two of the old runways form part of the private test track. The original line of one can be seen continuing across the field.**

LEFT **This is the main Boardroom of Ketteringham Hall as it now is. Clearly it was not nearly so opulent in the days when it was Badingham College.**

Postscript

In writing this book I have been caused to reflect at length over the time during which I knew Colin Chapman. As a man to work for I found him usually most satisfying. He delegated well and gave support when I needed it. I found him good at conveying to me what he wanted and explaining any background I might need to do my job. I also found him ready to listen to suggestions, not only from me and other collaborators, but from any source. I frequently found myself taking his part in office discussions; for example, on one occasion a draftsman had been visited by Chapman at his drawing board who had found much to criticise. After Chapman had left, the draftsman expressed his irritation to me, complaining that Chapman interfered too much in his work. While having some sympathy, I found myself saying "Look; if it doesn't have Chapman's mark on it it isn't a Lotus; and that's what we're here to create."

Having then praised Chapman in the workplace I have to add 'usually' or 'at times', since there were certainly occasions when I felt unable to judge how he would decide on a particular issue. I was consequently sometimes unable to progress the work without consulting him – I remember saying once "We'll have to wait to see which side of the bed he got out of today." At other times, for example when

I had to cope with an impatient visitor he would not come to see, I seethed with annoyance and frustration.

I should like to add a few words about Chapman the driver. On the race track he was outstanding and could bear comparison with any. He had to choose between driving and designing since to do both to the highest level was not practical. On the road his driving was like low flying. I often had cause to ride with him, mostly in his Jaguar or a Lotus Cortina. In some ways I felt safe with him even in the snow because he had such superb control. However, I was often scared at blind brows or other such hazards. I once asked him about this and he answered that the chances of meeting another like himself coming the other way were very small indeed. (By contrast, I also had the good fortune to ride with Jim Clark on the road in a Lotus Cortina. His driving was just as fast as Chapman's but less frightening since he did seem to slow down where I would have myself.)

Overall, to work for and with Colin Chapman was a very formative experience for a young man. His approach to problem solving, his enthusiasm and that genius for innovation are things which have left their mark, even if they have not rubbed off. I have never worked for another person whose words and deeds were to me so indelible.

Appendix – Factories

DURING HIS LIFETIME, Colin Chapman was involved in the setting-up and or creation of several factories; at Hornsey, north London, he established his first 'proper' workshops. Here, later, he was to add a first showroom. (see p.111)

From here, Lotus moved to Cheshunt in Hertfordshire. They took over a site in Delamare Road in 1959, occupying the new factory, consisting of two buildings, in June that year. The factory was operational from September, producing Elites in the Lotus Cars building which also housed the main offices, and racing and sports cars in the Lotus Components building, also home to Team Lotus. One of the decisions of the time was to build the Lotus Sevens on the first floor of Lotus Components. The building was provided with large sliding doors at each end of both floors, but no ramp or other vehicle access to the upper level. The problem was solved by the installation of a crane jib built into the roof structure. Every bare Lotus Seven body unit was hoisted up by this means and every finished chassis was lowered by it when ready for delivery. The present occupiers of the building have left the jib in place because it is so much a part of the structure that to remove it would be very costly.

About three years later the Lotus Cars building was extended rearwards into what had been the field where Elite bodies were stored. Again, to make the most of the space and the building costs an upper floor was installed. The ground floor became on one side the customer service stores and workshops and on the other the stores for the Lotus Cortina shop upstairs. The upper floor was again provided with a winch, but it was only for feeding in the painted and trimmed bodies received from Dagenham. The bodies arrived without the famous green stripe, which was added as the first step in the build process. Finished cars came down a ramp at the opposite end of the shop. When Chapman drove the first car down to signify the opening of the factory, he damaged it on the kerb of the ramp on the sharp bend at the top. Modifications were rapidly made that can be seen today.

The system within the Cortina shop was that the bodies would be put on three foot high trollies and be pushed from station to station round the building as the various parts were added. They were lifted from these only when complete and ready to be started for the first time. When it came to lifting a car down for the first time the intended means was not yet available. The problem was solved by raising it on two large garage jacks, one at each end, and then lowering it to the ground. During this manoeuvre the car was extremely unstable, but was got down safely despite white knuckles.

In 1966 the company moved again, this time to its present site at Hethel, Norfolk. Chapman involved himself in every detail of the design and layout, sometimes causing frustration for the architects Chaplin and Ferrant. The factory was built by Costains. As a company, Lotus made very great efforts to induce as many of their staff to move with them as possible. Finally no less than 50% made the move – something of a record. Chapman had taken on Ford's Dennis Austin to be Managing

OVERLEAF **The Lotus factory in Cheshunt – 1959. Motoring journalists were invited to visit the new Cheshunt Factory when it opened. This article was one of the results.**

The New Lotus Works

IMPRESSIVE NEW FACTORY AT CHESHUNT, HERTS.

ON Wednesday, 14 October, the new Lotus factory at Delamare Road, Cheshunt, Herts, was opened by Alderman Godfrey Lagden, MP, at a ceremony attended by over 200 guests. The assembled company included representatives of many of the firms and organisations associated with Lotus, a large number of racing drivers—most of them Lotus exponents—and several members of the motoring press.

DISTINGUISHED GATHERING

Among those present were L. P. Lee, W. L. T. Hassan and H. Spears of Coventry Climax, J. Sambrooke of Dunlop, F. Temple of the Ford Motor Co., J. Hawkins of the Firestone Tyre Co., G. B. Thomas and G. Murdoch of Esso, S. Henson of Ferodo, D. Herrick of Armstrong, G. Taylor of Girling, R. Wood and A. Cameron of Lucas, J. C. Smith, J. Ellis, F. Neville and J. Nuttall of Smiths Motor Accessories, L. Hands of Champion and R. White Smith and H. J. Lock of the Bristol Aircraft Co. Racing drivers in attendance included Cliff Allison, Innes Ireland, Alan Stacey, Peter Ashdown and Keith Greene.

The new factory consists of two similar buildings, separated by a central roadway, of which one, occupied by Lotus Cars, is devoted to Elite production while the other houses Lotus Components, Lotus Developments and the Service Department. Each building is of single span construction, roofs and walls are insulated, and there are no windows of normal type, their place being taken by transparent ventilators. Sliding doors, opening on to the central roadway, facilitate entry and exit for both cars and components.

In the north building, which is devoted solely to Elite production, the glass-fibre body/chassis units and all components are taken in at one end, and complete cars are driven out at the other. Construction work is divided into six stages, and the body/chassis units move along on low castor-wheeled trolleys which, in addition to permitting movement in any direction, also facilitate work underneath the cars. The capacity of this production line is at present upwards of 25 cars per week.

INTERNAL ORGANISATION

Lotus Cars is run by Colin Chapman as Director and General Manager, with Graham Lewis as Production Manager and Len Laverock as Works Manager. Before delivery all Elites are subjected to a searching road-test by Colin Chapman.

By contrast with the Elite shop, the south building is divided up by movable partitions into a number of separate sections. Just inside the main entrance are the offices and the Seven shop on the left, and the stores on the right. This section is run by Norman (Nobby) Clark as Director and General Manager, with Roy Badcock as Works Manager. In the centre of the building, occupying what amounts to a small cage, is the Developments section—Director, Mike Costin.

On the north side of this building are the Service Department, a panel-beating section (in which Seven bodywork is currently being turned out) and the sports/racing section—rather less busy at present than at any other time of the year. It has not yet proved possible to evolve a " production line " technique for sports/racing cars, but Seven chassis/body units are being " semi-mass-produced " at a rate of up to 15 a week.

AFTER-SALES SERVICE

The Service Department, which is equipped to deal with all the firm's models, is run by Tony Calder-Smith, and has already earned itself a very good reputation. During a recent visit an interesting selection of cars was found in this department—a Seventeen, Chris Barber's Elite, in for a routine service, with 13,000 miles on the distance recorder, and the Coventry - Climax - engined Seven which was taken to Goodwood for the Guild of Motoring Writers Motor Show Test Day.

Sales and administrative offices are housed in a separate block facing on to Delamare Road. Contemporary furnishings and decoration are widely in evidence, and the entrance hall features a large " tiled " Lotus badge in the centre of the floor, an open staircase to the first floor offices, and a variety of indoor plants. On the right is the showroom, on the left are the sales offices and above are the administrative offices and boardroom.

One great advantage of the new factory is that the site offers space for expansionsion, and to judge by the amount of work curently in progress it will not be long before the need for further floor space is felt. For the present, however, the chief impression of the majority of visitors is that at last the firm has a factory in keeping with its standing in the world of motoring.

ABOVE **The Lotus factory in Cheshunt in 1992. The main office building and part of the old Elite/Elan shop is occupied by Biccotest, making advanced electronic test gear for cabling. In this view the two windows to the left upstairs show where Colin Chapman had his office.**

ABOVE **From the inside, Chapman's office is little changed but for the absence of car models and the presence of a computer. The office is now used by Derek Henderson, Director & General Manager of Biccotest.**

Director and he was charged with the supervision of the factory preparation and move. In the event the whole move took place over a weekend and the first Hethel-produced car came from the factory before the end of the Monday.

Part of the attraction of Hethel to Chapman, was, of course the fact that it had been an airfield. Lotus bought about half of the original war time base, the rest being returned to the farmers who had owned it before the war. Chapman thus had at his disposal a runway and other roads which became the car test track but doubled to serve him and other pilots in the company and their visitors.

A further establishment that Chapman had a personal interest in was, of course, Ketteringham Hall, where Team Lotus has its home. This building, parts of which date from Tudor times, was bought by Chapman in the dilapidated state into which it had fallen when it had been Badingham College, a boys' school for 13- to 18-year-olds. Badingham College was originally in Leatherhead in Surrey, but, following the death of the founder and a family disagreement, was moved to Ketteringham by his son, the Rev Alan Wilkie in 1965 with greatly reduced resources. Already, from 1950, there had been a preparatory school there,

ABOVE **The Lotus Cortina extension is at the time of writing occupied partly by Biccotest and Electrocircuits (upper floor) and by Power Development Ltd (ground floor). This was the door through which the bodies were hoisted to start their movement through the assembly shop. Below the entrance to the assembly floor was the stores entrance – now bricked up.**

RIGHT **Power Development Ltd have added many ducts to ventilate the chemical processes they use.**

and Wilkie bought the house from the owners for £45000.

The school did not thrive and gradually funds became so short that non-teaching staff were cut right down; Mrs Wilkie did the cooking and the boys did the washing-up by rota. At the final demise in 1969, Chapman bought it, paying only £39000. It was not until 1976 that Team Lotus was moved there. To start with it was used by Lotus as a Field Service training school, which was run by Jim Endruweit, who had been the Chief Mechanic of Team Lotus during the Cheshunt days. Ketteringham Hall went on to house Chapman's 'think tank' until its occupation by the team proper.

Completed Lotus Cortinas descended this ramp for road test and despatch.

Chapman's alteration to the kerb of the ramp can still be seen.

The rear part of the Lotus Cortina floor housed the design & development area for Lotus Cars. It is now occupied by Electrocircuits (PC) Ltd. From their exit onto the ramp they have this view of the rear of the Lotus Components building. It was in here that the Lotus 25 was designed and developed. The left of the building was occupied by Team Lotus.

From the front the Components building is still almost unchanged. It is used by Omega furniture, who make a range of lounge furniture. Beneath the Sale sign was the entrance to the spare parts counter.

This is the hoist by which Lotus Sevens were moved in and out of the assembly area.

The Lotus Seven assembly shop now houses Omega's furniture showroom.

Where Lotus Components once made sports and racing cars for sale, Omega now manufacture their furniture. One model in their range is called Lotus.

Workshops once used by Team Lotus are now the woodwork shops for Omega.

Index